Lecture Notes in Computer Science

Lecture Notes in Bioinformatics 13954

The series Lecture Notes in Bioinformatics (LNBI) was established in 2003 as a topical subseries of LNCS devoted to bioinformatics and computational biology.

The series publishes state-of-the-art research results at a high level. As with the LNCS mother series, the mission of the series is to serve the international R & D community by providing an invaluable service, mainly focused on the publication of conference and workshop proceedings and postproceedings.

Marcelo S. Reis · Raquel C. de Melo-Minardi
Editors

Advances in Bioinformatics and Computational Biology

16th Brazilian Symposium on Bioinformatics, BSB 2023
Curitiba, Brazil, June 13–16, 2023
Proceedings

 Springer

Editors
Marcelo S. Reis (ID)
Universidade Estadual de Campinas
Campinas, Brazil

Raquel C. de Melo-Minardi (ID)
Universidade Federal de Minas Gerais
Belo Horizonte, Brazil

ISSN 0302-9743   ISSN 1611-3349 (electronic)
Lecture Notes in Bioinformatics
ISBN 978-3-031-42714-5   ISBN 978-3-031-42715-2 (eBook)
https://doi.org/10.1007/978-3-031-42715-2

LNCS Sublibrary: SL8 – Bioinformatics

This Springer imprint is published by the registered company Springer Nature Switzerland AG
The registered company address is: Gewerbestrasse 11, 6330 Cham, Switzerland

Paper in this product is recyclable.

Preface

The Brazilian Symposium on Bioinformatics (BSB) is an international, annual scientific conference with focus on Bioinformatics, Computational Biology, Systems Biology, and related areas. It is organized by the Brazilian Computer Society (*Sociedade Brasileira de Computação* – SBC), through the Special Committee for Computational Biology (*Comissão Especial de Biologia Computacional* – CE-BioComp), which is presently coordinated by Raquel C. de Melo-Minardi (UFMG) and co-coordinated by Kele Belloze (CEFET/RJ). BSB 2023 was the 16th edition of the conference, and it was held during June 13–16, 2023, at the FIEP Event Center, Curitiba, Brazil. Curitiba is the capital of Paraná (PR), a state in southern Brazil.

The BSB 2023 conference had as general chairs Fabricio Martins Lopes (UTFPR) and Alexandre R. Paschoal (UTFPR). The organization committee had as members Dieval Guizelini (UFPR) and Roberto Tadeu Raittz (UFPR). The international Program Committee this year was composed of 37 members from Brazil and also from Canada, France, Germany, Mexico, and Uruguay. The conference, which accepted contributions in the form of full and short papers, received a total of 24 submissions, with 14 works being accepted (13 full papers and 1 short paper). The submitted works passed through a single-blind review process, with each paper having at least three independent reviews. Each of the 14 accepted papers is included in this collection and was presented at the conference by one of its authors, at one of the three technical sessions that were held at BSB 2023.

The 16th edition of BSB was co-located with the 19th International Congress of the Brazilian Association of Bioinformatics and Computational Biology (*Associação Brasileira de Bioinformática e Biologia Computacional* – AB3C). That congress, also known as X-Meeting, is one of the two major Brazilian Bioinformatics conferences, the other one being BSB itself. While BSB 2023 accepted full and short papers, X-Meeting 2023 focused on poster submissions. Therefore, the co-located event joined the communities of the two conferences into a single, synergistic event, which also allowed the realization of one of the largest Bioinformatics conferences ever held in Brazil, with the participation of over 500 delegates! The co-located event also shared the keynote speakers, with lectures by Robert Finn (EMBL-EBI, UK), Martin Morgan (Fred Hutchinson Cancer Research Center, USA), Miguel Rocha (Universidade do Minho, Portugal), Helder Nakaya (Hospital Israelita Albert Einstein, Brazil), Sameer Velankar (EMBL-EBI, UK), and João Meidanis (UNICAMP, Brazil); this latter speaker was also honored by the Conference Organization, for his contributions to the development of Bioinformatics in Brazil.

Finally, we thank a lot all the people that made BSB + X-Meeting 2023 possible: the Program Committee members, who accomplished dozens of reviews in a tight schedule; the organization chairs, committees and volunteers (from both BSB and X-Meeting sides), who assured the realization of the co-located event from a logistical standpoint; the support of the AB3C and the financial support of Fundação Araucária; the partnership with Springer for, once more, publishing the BSB proceedings; the keynote speakers for accepting the invitation for their talks; the instructors of the two mini-courses offered during BSB 2023; and last but not least, all the authors that sent their contributions to the 16th edition of BSB. To all of them: thank you! (*muito obrigado!*)

June 2023

Marcelo S. Reis
Raquel C. de Melo-Minardi

Organization

General Chairs

Fabricio Martins Lopes   Universidade Tecnológica Federal do Paraná, Brazil

Alexandre R. Paschoal   Universidade Tecnológica Federal do Paraná, Brazil

Organization Committee

Dieval Guizelini   Universidade Federal do Paraná, Brazil

Roberto Tadeu Raittz   Universidade Federal do Paraná, Brazil

Program Committee Chairs

Marcelo S. Reis   Universidade Estadual de Campinas, Brazil

Raquel C. de Melo-Minardi   Universidade Federal de Minas Gerais, Brazil

Steering Committee (CE-BioComp)

Raquel C. de Melo-Minardi   Universidade Federal de Minas Gerais, Brazil

Kele Belloze   Centro Federal de Educação Tecnológica Celso Suckow da Fonseca, Brazil

Sérgio Lifschitz   Pontifícia Universidade Católica do Rio de Janeiro, Brazil

Waldeyr Mendes Cordeiro da Silva   Imprensa Nacional, Brazil

Fernanda Nascimento Almeida   Universidade Federal do Grande ABC, Brazil

Program Committee

Said Adi   Universidade Federal do Mato Grosso do Sul, Brazil

Yalbi Balderas-Martinez   Instituto Nacional de Enfermedades Respiratorias Ismael Cosío Villegas, Mexico

David Correa Martins, Jr.	Universidade Federal do ABC, Brazil
Luís Cunha	Universidade Federal Fluminense, Brazil
Carlos da Silveira	Universidade Federal de Itajubá, Brazil
Daniel de Oliveira	Universidade Federal Fluminense, Brazil
Marcilio de Souto	Université d'Orléans, France
Zanoni Dias	Universidade Estadual de Campinas, Brazil
Luciano Digiampietri	Universidade de São Paulo, Brazil
Karina dos Santos Machado	Universidade Federal do Rio Grande, Brazil
André Fujita	Universidade de São Paulo, Brazil
Thaís Gaudêncio do Rêgo	Universidade Federal da Paraíba, Brazil
Ana Carolina Guimarães	Fundação Oswaldo Cruz, Brazil
Christian Höner zu Siederdissen	Friedrich-Schiller-Universität Jena, Germany
Seyed Jamalaldin Haddadi	Universidade Estadual de Campinas, Brazil
Andre Kashiwabara	Universidade Tecnológica Federal do Paraná, Brazil
Manuel Lafond	University of Sherbrooke, Canada
Giuseppe Leite	Universidade Federal de São Paulo, Brazil
Felipe Louza	Universidade Federal de Uberlândia, Brazil
Ariane Machado-Lima	Universidade de São Paulo, Brazil
Fabrício Martins Lopes	Universidade Tecnológica Federal do Paraná, Brazil
Milton Nishiyama, Jr.	Instituto Butantan, Brazil
Sergio Pantano	Institut Pasteur de Montevideo, Uruguay
Thiago Parente	Fundação Oswaldo Cruz, Brazil
Mariana Recamonde-Mendoza	Universidade Federal do Rio Grande do Sul, Brazil
Luiz Manoel Rocha Gadelha, Jr.	Laboratório Nacional de Computação Científica, Brazil
Danilo Sanches	Universidade Tecnológica Federal do Paraná, Brazil
Leandro Santos	Instituto Nacional de Câncer, Brazil
João Carlos Setubal	Universidade de São Paulo, Brazil
Sabrina Silveira	Universidade Federal de Viçosa, Brazil
Guilherme Telles	Universidade Estadual de Campinas, Brazil
Nayara Toledo	Instituto Nacional de Câncer, Brazil
Diogo Tschoeke	Universidade Federal do Rio de Janeiro, Brazil
Alfredo Varela	Universidad Nacional Autónoma de México, Mexico
Didier Vega-Oliveros	Universidade Estadual de Campinas, Brazil
Glauber Wagner	Universidade Federal de Santa Catarina, Brazil
Adriano V. Werhli	Universidade Federal do Rio Grande, Brazil

Realization

Sociedade Brasileira de Computação (SBC), Brazil

Partner Association

Associação Brasileira de Bioinformática e Biologia Computacional (AB3C), Brazil

Financial Support

Fundação Araucária, Brazil

Institutional Organization

Universidade Tecnológica Federal do Paraná, Brazil

Universidade Federal do Paraná, Brazil

Contents

Block Interchange and Reversal Distance
on Unbalanced Genomes

Alexsandro Oliveira Alexandrino[1]([✉]), Gabriel Siqueira[1], Klairton Lima Brito[1],
Andre Rodrigues Oliveira[2], Ulisses Dias[3], and Zanoni Dias[1]

[1] Institute of Computing, University of Campinas, Campinas, Brazil
{alexsandro,gabriel.siqueira,klairton,zanoni}@ic.unicamp.br
[2] Computing and Informatics Department, Mackenzie Presbyterian University,
São Paulo, Brazil
andre.rodrigues@mackenzie.br
[3] School of Technology, University of Campinas, Campinas, Brazil
ulisses@ft.unicamp.br

Abstract. One method for inferring the evolutionary distance between
two organisms is to find the *rearrangement distance*, which is defined as
the minimum number of genome rearrangements required to transform
one genome into the other. Rearrangements that do not alter the genome
content are known as conservative. Examples of such rearrangements
include: *reversal*, which reverts a segment of the genome; *transposition*,
which exchanges two consecutive blocks; *block interchange (BI)*, which
exchanges two blocks at any position in the genome; and *double cut and
join (DCJ)*, which cuts two different pairs of adjacent blocks and joins
them in a different manner. Initially, works in this area involved comparing genomes that shared the same set of conserved blocks. Nowadays,
researchers are investigating unbalanced genomes (genomes with a distinct set of genes), which requires the use of non-conservative rearrangements such as *insertions* and *deletions* (*indels*). In cases where there are
no repeated blocks and the genomes have the same set of blocks, the BI
Distance and the Reversal Distance have polynomial-time algorithms,
while the complexity of the BI and Reversal Distance problem remains
unknown. In this study, we investigate the BI and Indel Distance and the
BI, Reversal, and Indel Distance on genomes with different gene content
and no repeated genes. We present 2-approximation algorithms for each
problem using a variant of the breakpoint graph structure.

Keywords: Block Interchange · Reversal · Unbalanced Genomes

This work was supported by the National Council of Technological and Scientific Development, CNPq (grants 140272/2020-8, 202292/2020-7 and 425340/2016-3), the Coordenação de Aperfeiçoamento de Pessoal de Nível Superior - Brasil (CAPES) - Finance
Code 001, and the São Paulo Research Foundation, FAPESP (grants 2013/08293-7,
2015/11937-9, 2019/27331-3, 2021/13824-8, and 2022/13555-0).

M. S. Reis and R. C. de Melo-Minardi (Eds.): BSB 2023, LNBI 13954, pp. 1–13, 2023.
https://doi.org/10.1007/978-3-031-42715-2_1

1 Introduction

Mutations play a significant role during the evolutionary process. When these mutations affect large stretches of a genome, they are called *genome rearrangements*. By analyzing the relative order of genes in genomes of related species, we can compute a sequence of rearrangements that transforms one genome into another. Based on the principle of parsimony, the scenario with the least number of rearrangements is assumed to be the most likely to have occurred.

The problem of finding the minimum number of rearrangements required to transform one genome into another, known as the rearrangement distance, is addressed using a model that defines which rearrangements should be considered. There are several genome rearrangement models, including conservative and non-conservative events. Conservative events, such as reversal, block interchange (BI), transposition, and double cut and join (DCJ), do not alter the amount of genetic material. In contrast, non-conservative events, such as insertion and deletion, add or remove genetic material at specific positions in the genome.

The computation of the rearrangement distance between two genomes can be accomplished in polynomial time for certain models, while for others, it is NP-hard. This depends on the level of information available, such as the orientation of genes in each genome. When gene orientations are considered, both the Reversal Distance and the DCJ Distance can be solved in polynomial time [12, 16]. However, when orientations are not known, these distances become NP-hard, as demonstrated by previous studies [7, 8, 13].

Since block interchanges and transpositions change only the relative position of elements but not their orientations, they do not consider gene orientation [11]. The Block Interchange Distance has an exact polynomial time algorithm [9], while the Transposition Distance is NP-hard [6].

The literature on genome rearrangements started the study of the distance between unbalanced genomes (genomes with a distinct set of genes) in 2000 [10], and most of the models use *indels*, which refers to both insertions and deletions. Considering gene orientation, the DCJ and Indel Distance [5] and the Reversal and Indel Distance [15] are both solvable in polynomial time, while the Transposition and Indel Distance is NP-hard [1, 2].

Here we study the Block Interchange and Indel Distance and the Block Interchange, Reversal, and Indel Distance, considering that genomes have a distinct set of genes, but there are no occurrences of repeated genes in a genome. We present lower bounds and 2-approximation algorithms for these problems.

2 Definitions

An instance for a rearrangement distance problem has a source genome $\mathcal{G}_1$ and a target genome $\mathcal{G}_2$. We represent the target genome $\mathcal{G}_2$ with the identity string $\iota^n = (+1\ +2\ \ldots\ +n)$, where each element ι_i^n maps a gene or a maximal continuous sequence of genes without correspondence in $\mathcal{G}_1$. We say that $1, 2, \ldots, n$

(without signs) are *labels*. We represent the source genome $\mathcal{G}_1$ with a string $A = (A_1 \ A_2 \ \dots \ A_m)$, where A_i maps a gene, using the same mapping of labels and genes used for the target genome, or it represents a maximal continuous sequence of genes without correspondence in $\mathcal{G}_2$. If A_i maps a gene of $\mathcal{G}_1$, then it has a "+" sign if the gene with same label in $\mathcal{G}_2$ has the same orientation, and it has a "−" sign otherwise. For any element A_i that maps a continuous sequence of genes without correspondence in $\mathcal{G}_2$, we set $A_i = \alpha$ without any sign, since this element will be removed regardless of its content.

We use $-A_i$ to denote the element A_i with its orientation reversed. For example, if $A_i = -1$, then $-A_1 = +1$. For the models where gene orientation is not considered, as the Block Interchange and Indel Distance, we can just omit the signs or consider that every element has a "+" sign.

The *alphabet* Σ_σ of a string σ is the set of labels present in σ. Note that $\Sigma_A \setminus \Sigma_{\iota^n} = \{\alpha\}$. Furthermore, there are no adjacent elements in ι^n such that both of them belong to $\Sigma_{\iota^n} \setminus \Sigma_A$, since any maximal continuous segment of genes without correspondence in $\mathcal{G}_1$ are mapped into a single element in ι^n. For the strings $A = (+6 \ \alpha \ -3 \ +4 \ +1 \ \alpha)$ and $\iota^6 = (+1 \ +2 \ +3 \ +4 \ +5 \ +6)$, we have $\Sigma_A \cap \Sigma_{\iota^6} = \{1,3,4,6\}$, $\Sigma_A \setminus \Sigma_{\iota^6} = \{\alpha\}$, $\Sigma_{\iota^6} \setminus \Sigma_A = \{2,5\}$.

Given a string A with $|A| = m$, a block interchange $\mathcal{BI}(i,j,k,l)$, with $1 \leq i \leq j < k \leq l \leq m$, is a rearrangement that acts on the segments $(A_i \ \dots \ A_j)$ and $(A_k \ \dots \ A_l)$ generating the string $A \cdot \mathcal{BI}(i,j,k,l) = (A_1 \ \dots \ A_{i-1} \ \underline{A_k \ \dots \ A_l} \ A_{j+1} \ \dots \ A_{k-1} \ \underline{A_i \ \dots \ A_j} \ A_{l+1} \ \dots \ A_m)$.

Given a string A with $|A| = m$, a reversal $\rho(i,j)$, with $1 \leq i \leq j \leq m$, inverts the segment $(A_i \ \dots \ A_j)$ and changes the orientation of the elements in it. It generates the string $A \cdot \rho(i,j) = (A_1 \ \dots \ A_{i-1} \ \underline{-A_j \ \dots \ -A_i} \ A_{j+1} \ \dots \ A_m)$.

Given a string A with $|A| = m$, an insertion $\phi(i,S)$, where $0 \leq i \leq m$ and S is a string, is a rearrangement which inserts S in the position $i+1$ of a string. When applied to A, we have $A \cdot \phi(i,S) = (A_1 \ \dots \ A_i \ \underline{S_1 \ \dots \ S_{|S|}} \ A_{i+1} \ \dots \ A_m)$.

Given a string A with $|A| = m$, a deletion $\psi(i,j)$, with $1 \leq i \leq j \leq m$, removes the segment $(A_i \ \dots \ A_j)$ from the string A. When applied to A, we have $A \cdot \psi(i,j) = (A_1 \ \dots \ A_{i-1} \ A_{j+1} \ \dots \ A_m)$.

A rearrangement model $\mathcal{M}$ defines the set of allowed rearrangements to compute the distance in a rearrangement distance problem. Given an instance (A, ι^n), the distance $d_{\mathcal{M}}(A, \iota^n)$ is the minimum number of operations in $\mathcal{M}$ that transforms A into ι^n. Since both models studied in this paper have indels, we chose not to mention it in the model acronym, so we use $d_{\mathcal{BI}}(A, \iota^n)$ and $d_{\rho,\mathcal{BI}}(A, \iota^n)$ for the Block Interchange and Indel Distance, and the Block Interchange, Reversal, and Indel Distance, respectively.

2.1 Labeled Cycle Graph

The Labeled Cycle Graph [2,14] is an adaptation of the breakpoint graph and the cycle graph created to deal with unbalanced genomes.

Given an instance (A, ι^n), we create the strings $\pi^A = (\pi_1^A \ \dots \ \pi_{n'}^A)$ and $\pi^\iota = (\pi_1^\iota \ \dots \ \pi_{n'}^\iota)$ as copies of A and ι^n, respectively, but removing elements that do not belong to the set $\Sigma_A \cap \Sigma_{\iota^n}$. We extend both strings by adding the elements $\pi_0^A = 0$, $\pi_0^\iota = 0$, $\pi_{n'+1}^A = n+1$, and $\pi_{n'+1}^\iota = n+1$. We use $|\pi^A| = |\Sigma_A \cap \Sigma_{\iota^n}| = n'$ to denote the size of these strings without considering the extended elements 0 and $n+1$.

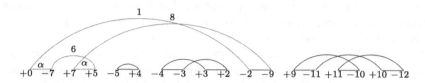

Fig. 1. Labeled Cycle Graph for the strings $A = (\alpha +7\ \alpha -5\ -4\ +3\ -2\ +9\ +11\ +10)$ and ι^n, with $n = 11$. There are four cycles in this graph. The cycle $C_1 = (6, 1, 2)$ is a divergent cycle with $\Lambda(C_1) = 4$. All the other cycles have 0 runs. The cycle $C_2 = (3)$ is a trivial cycle. The cycle $C_3 = (5, 4)$ is a divergent cycle. The cycle $C_4 = (9, 7, 8)$ is an oriented cycle.

The Labeled Cycle Graph for (A, ι^n) is the undirected graph $G(A, \iota^n) = (V, E, \ell)$, where $V = \{+\pi_0^A, -\pi_1^A, +\pi_1^A, -\pi_2^A, +\pi_2^A, \ldots, -\pi_{n'}^A, +\pi_{n'}^A, -\pi_{n'+1}^A\}$ is the set of vertices; $E = E_s \cup E_t$ is the set of edges, which is divided into source (E_s) and target (E_t) edges; and ℓ is an edge labeling function.

Source edges connect vertices that are adjacent in π^A, while target edges connect vertices that are adjacent in π^ι. The set of source edges $E_s = \{e_i = (+\pi_{i-1}^A, -\pi_i^A) : 1 \le i \le n'+1\}$. A source edge $e_i = (+\pi_{i-1}^A, -\pi_i^A)$ has *index* i. The label $\ell(e_i) = \emptyset$ if π_{i-1}^A and π_i^A are consecutive in A. Otherwise, we have $\ell(e_i) = \alpha$. The set of target edges $E_t = \{e_i^t = (+\pi_{i-1}^\iota, -\pi_i^\iota) : 1 \le i \le n'+1\}$. A target edge $e_i^t = (+\pi_{i-1}^\iota, -\pi_i^\iota)$ has *index* i. The label $\ell(e_i^t) = \emptyset$ if π_{i-1}^ι and π_i^ι are consecutive. Otherwise, the label $\ell(e_i^t) = \pi_{i-1}^\iota + 1$.

We say that an edge is clean if it has empty label; otherwise, we say that the edge is labeled.

Since there are exactly one source and one target edge incident to each vertex, there exists a unique decomposition of the graph into a collection of edge alternating cycles.

We draw the graph by arranging the vertices horizontally, following the order in which they appear in π^A. The source edges are displayed as horizontal lines while the target edges are shown as arcs. Edges that have a label are marked in red, and the label is placed above the edge. Figure 1 provides an illustration of this representation.

Each cycle C in $G(A, \iota^n)$ is denoted by the list of source edges indices that belong to C. For a cycle $C = (c_1, c_2, \ldots, c_k)$, we construct the list of indices starting with the rightmost source edge (i.e., $c_1 > c_i$, for all $1 < i \le k$) and traversing it from right to left.

A cycle with k source edges is called a k-cycle. A 1-cycle is called trivial. The number of cycles in $G(A, \iota^n)$ is denoted by $c(A, \iota^n)$. For a rearrangement β, we define $\Delta c(A, \iota^n, \beta) = (|\pi^A| + 1 - c(A, \iota^n)) - (|\pi^A \cdot \beta| + 1 - c(A \cdot \beta, \iota^n))$.

A cycle $C = (c_1, c_2, \ldots, c_k)$ is *oriented* if the values $(c_1, c_2, \ldots, c_k)$ do not form a decreasing sequence. Given a cycle $C = (c_1, c_2, \ldots, c_k)$, a source edge e_{c_i} is *convergent* if it is traversed from right to left, and it is *divergent* otherwise. Note that e_{c_1} is always convergent by our convention of how the cycle is traversed when listing indices. A pair of edges (c_i, c_j) is divergent if one of the source edges is divergent and the other is convergent. A cycle is divergent if it has at least one divergent source edge, and it is convergent otherwise.

A graph $G(A, \iota^n)$ has divergent cycles if, and only if, A has at least one element with "$-$" sign. Therefore, for the Block Interchange and Indel Distance there are only convergent cycles in the graph.

An *insertion run* is a maximal path that starts and ends with labeled target edges and has no labeled source edge. Similarly, a *deletion run* is a maximal path that starts and ends with labeled source edges and has no labeled target edge. The number of runs in a cycle C is given by $\Lambda(C)$.

The *indel potential* of a cycle C is a value of how much indels are necessary to turn $\Lambda(C) = 0$ without merging cycles or creating new cycles with runs in it. We define the indel potential of C as follows:

$$\lambda(C) = \begin{cases} \left\lceil \frac{\Lambda(C)+1}{2} \right\rceil, & \text{if } \Lambda(C) > 0 \\ 0, & \text{otherwise.} \end{cases}$$

We also denote by $\lambda(A, \iota^n)$ the sum of indel potentials of all cycles in $G(A, \iota^n)$, that is, $\lambda(A, \iota^n) = \sum_{C \in G(A, \iota^n)} \lambda(C)$. We also have that $\Delta\lambda(A, \iota^n, \beta) = \lambda(A, \iota^n) - \lambda(A \cdot \beta, \iota^n)$, which denotes the change in the indel potential of the graph caused by a rearrangement.

Lemma 1 (Alexandrino et al. [2]). *For any deletion ψ and strings A and ι^n, we have that $\Delta c(A, \iota^n, \psi) + \Delta\lambda(A, \iota^n, \psi) \leq 1$.*

Lemma 2 (Alexandrino et al. [2]). *For any insertion ϕ and strings A and ι^n, we have that $\Delta c(A, \iota^n, \phi) + \Delta\lambda(A, \iota^n, \phi) \leq 1$.*

Lemma 3. *For any block interchange $\mathcal{BI}$ and strings A and ι^n, we have that $\Delta c(A, \iota^n, \mathcal{BI}) + \Delta\lambda(A, \iota^n, \mathcal{BI}) \leq 2$.*

Proof. We divide this proof according to the number of cycles affected by $\mathcal{BI}$ [9].

If $\mathcal{BI}$ affects four cycles C_1, C_2, C_3, and C_4, then it merges these cycles into two new cycles C_1' and C_2'. In the best scenario, two deletion runs and two insertion runs from C_1 and C_3 are merged in C_1'. Similarly, two deletion runs and two insertion runs from C_2 and C_4 are merged in C_2'. In this case, $\Lambda(C_1') = \Lambda(C_1) + \Lambda(C_3) - 2$ and $\Lambda(C_2') = \Lambda(C_2) + \Lambda(C_4) - 2$. Therefore, $\Delta\lambda(A, \iota^n, \mathcal{BI}) = 4$ and $\Delta c(A, \iota^n, \mathcal{BI}) + \Delta\lambda(A, \iota^n, \mathcal{BI}) = 2$. An example is presented in Fig. 2.

If $\mathcal{BI}$ affects three cycles C_1, C_2 and C_3, then it merges these cycles into a new cycle C'. Similarly to the previous case, in the best scenario, the number of runs decreases in four and $\Lambda(C') = \Lambda(C_1) + \Lambda(C_2) + \Lambda(C_3) - 4$. Therefore, $\Delta\lambda(A, \iota^n, \mathcal{BI}) = 4$ and $\Delta c(A, \iota^n, \mathcal{BI}) + \Delta\lambda(A, \iota^n, \mathcal{BI}) = 2$.

If $\mathcal{BI}$ affects two cycles C_1 and C_2, then it turns these cycles into two new cycles or into four new cycles. If it turns C_1 and C_2 into two new cycles C_1' and C_2', then, in the best scenario, the number of runs decreases in four and $\Lambda(C_1') = \Lambda(C_1) - 2$ and $\Lambda(C_2') = \Lambda(C_1) - 2$. Therefore, the indel potential decreases by one for each cycle, so $\Delta\lambda(A, \iota^n, \mathcal{BI}) = 2$, and $\Delta c(A, \iota^n, \mathcal{BI}) = 0$.

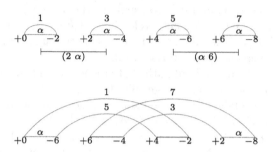

Fig. 2. Example of a block interchange that acts on four cycles. In this example, we have $A = (0\ \alpha\ 2\ \alpha\ 4\ \alpha\ 6\ \alpha\ 8)$ and $n = 7$. The indel potential of the original graph is equal to $4 \times \lceil(2+1)/2\rceil = 8$ and the indel potential of the new graph is equal to $\lceil(2+1)/2\rceil + \lceil(2+1)/2\rceil = 4$.

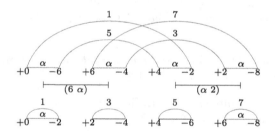

Fig. 3. Example of a block interchange that acts on two cycles creating four new cycles. In this example, we have $A = (0\ \alpha\ 6\ \alpha\ 4\ \alpha\ 2\ \alpha\ 8)$ and $n = 7$. The indel potential of the original graph is equal to $\lceil(4+1)/2\rceil + \lceil(4+1)/2\rceil = 6$ and the indel potential of the new graph is equal to $\lceil(2+1)/2\rceil + \lceil(1+1)/2\rceil + \lceil(1+1)/2\rceil + \lceil(2+1)/2\rceil = 6$.

If it affects two cycles C_1 and C_2, and it turns these cycles into four new cycles C_1', C_2', C_3', and C_4', then, in the best scenario, two pairs of deletion runs are merged, but note that each cycle has at least one insertion run, as shown in Fig. 3. So, $\Lambda(C_1') = X$, such that $1 \leq X < \Lambda(C_1)$, $\Lambda(C_2') = min(\Lambda(C_1) - X - 2, 1)$, $\Lambda(C_3') = Y$, such that $1 \leq Y < \Lambda(C_2)$, and $\Lambda(C_4') = min(\Lambda(C_2) - Y - 2, 1)$. Therefore, the indel potential of the graph remains the same.

If $\mathcal{BI}$ affects one cycle C_1, then it turns this cycle into a new cycle or into three new cycles. If it does not change the number of cycles, then, in the best scenario, it can decrease the number of runs in the cycle by four and $\Delta\lambda(A, \iota^n, \mathcal{BI}) = 2$. If it turns this cycle into three new cycles C'_1, C'_2, and C'_3, then, in the best scenario, two pairs of deletion runs are merged, but note that each cycle has at least one insertion run. Similarly to the previous case, the indel potential of the graph remains the same.   □

Lemma 4 (Willing et al. [14]). *For any reversal ρ and strings A and ι^n, we have that $\Delta c(A, \iota^n, \rho) + \Delta\lambda(A, \iota^n, \rho) \leq 1$.*

The graph $G(A, \iota^n)$ has only trivial cycles and indel potential of zero if, and only if, the strings $A = \iota^n$. Note that, when $A = \iota^n$, we have $|\pi^A| + 1 - c(A, \iota^n) + \lambda(A, \iota^n) = 0$.

Lemma 5. *For any strings A and ι^n, we have*

$$d_{\mathcal{BI}}(A, \iota^n) \geq \left\lceil \frac{|\pi^A| + 1 - c(A, \iota^n) + \lambda(A, \iota^n)}{2} \right\rceil.$$

Proof. Since $|\pi^{A'}| + 1 - c(A', \iota^n) + \lambda(A', \iota^n) = 0$ only if $A' = \iota^n$, a sequence of rearrangements that transform A into ι^n must decrease the value of $|\pi^A| + 1 - c(A, \iota^n) + \lambda(A, \iota^n)$ to zero. From Lemmas 1 to 3, a rearrangement can decrease this value by at most two and, therefore, the bound follows.   □

Lemma 6. *For any strings A and ι^n, we have*

$$d_{\rho,\mathcal{BI}}(A, \iota^n) \geq \left\lceil \frac{|\pi^A| + 1 - c(A, \iota^n) + \lambda(A, \iota^n)}{2} \right\rceil.$$

Proof. Similar to the proof of Lemma 5 considering Lemmas 1 to 4.   □

3   2-Approximation Algorithms for the Distance Problems

In this section, we introduce algorithms with approximation factors of 2 that use the graph structure presented in the previous section. Alexandrino et al. [2] presented a result on how to remove insertion runs from the graph and decrease the indel potential, but only considering unsigned strings. We present how this can be done for signed strings as well.

Lemma 7. *For any strings A and ι^n, if $G(A, \iota^n)$ has insertion runs, then there exists an insertion ϕ with $\Delta c(A, \iota^n, \phi) + \Delta\lambda(c, \iota^n, \phi) = 1$.*

Proof. Consider the insertion run $(v_1, v_2, \ldots, v_j)$ of a cycle C, such that v_1 has the same sign as the element of A that corresponds to v_1 and (v_1, v_2) is a labeled target edge. Let $o_1, o_2, \ldots, o_k$ be indices such that (v_{o_i}, v_{o_i+1}) is the i-th labeled target edge of this run.

We construct $S = (x_1, x_2, \ldots, x_k)$ as follows: for $1 \leq i \leq k$, if v_{o_i+1} has a "$-$" sign, then $x_i = \ell((v_{o_i}, v_{o_i+1}))$; otherwise, $x_i = -\ell((v_{o_i}, v_{o_i+1}))$. The insertion of S after the element of A corresponding to v_1 removes the run and adds k cycles in the graph. A trivial cycle is created with the vertices $(v_1, -x_1)$. For each element x_i, with $1 \leq i < k$, there is a cycle $(+x_i, v_{o_i+1}, v_{o_i+2}, \ldots, v_{o_{i+1}}, -x_{i+1}, +x_i)$. The last vertex $+x_k$ belongs to what is left of the cycle C or to a trivial cycle, in the case where all target edges of C belong to the removed run. An example of this operation is shown in Fig. 4.

If $\Lambda(C) \leq 2$, then removing a run of C reduces both the number of runs and the indel potential of the graph by one. Otherwise, removing an insertion run leads to the merging of two deletion runs. In this case, the number of runs of C decreases by two and the indel potential of the graph decreases by one. As the insertion adds k elements in A and k cycles in the graph with no runs, we have $\Delta c(A, \iota^n, \phi) = 0$. Therefore, $\Delta c(A, \iota^n, \phi) + \Delta \lambda(c, \iota^n, \phi) = 1$. $\square$

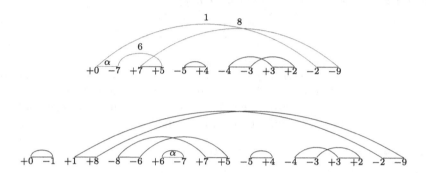

Fig. 4. Example of a insertion that removes a run from a cycle. In this example, we have the insertion run $(+0, -2, -9, +7, +5, -7)$. The insertion of $(+1\ -8\ +6)$ at the start of A removes this run and creates three new cycles.

Now, we show how block interchange operations can be used to increase the number of cycles in the graph without increasing the indel potential.

Lemma 8. *For any strings A and ι^n, such that $|\pi^A| + 1 - c(A, \iota^n) > 0$ and $G(A, \iota^n)$ has no labeled target edges, there exists a block interchange $\mathcal{BI}$ such that $\Delta c(A, \iota^n, \mathcal{BI}) + \Delta \lambda(c, \iota^n, \mathcal{BI}) = 2$.*

Proof. Consider that $G(A, \iota^n)$ has an oriented cycle $C = (c_1, \ldots, c_\ell)$, and let c_i, c_j, c_k be a triple such that $i < j < k$ and $c_i > c_k > c_j$. Such triple always exists in an oriented cycle and it is called an oriented triple [4]. A block interchange applied on these three source edges creates three cycles C', C'', and C''' [4]. Let S_1 and S_2 be the two segments changed by the block interchange. If the source edges are labeled, we can move the elements α in such a way that they end up in the same cycle. To do this, we include the segment to be removed from the first source edge in S_1 and the segment to be removed from the third source edge in S_2. In this way, the segments to be removed are merged in a single source edge, as shown in Fig. 5. An analogous operation is used if only two of these source edges are labeled. Therefore, this block interchange does not affect the indel potential of the graph and increases the number of cycles by two.

Consider that $G(A, \iota^n)$ has only non-oriented cycles and let $C = (c_1, \ldots, c_\ell)$ be a cycle of $G(A, \iota^n)$. Bafna and Pevzner [4] showed that for every e_{c_i} and e_{c_j} from C, with $c_i > c_j$, there exists a cycle $D = (d_1, \ldots, d_{\ell'})$ with source edges e_{d_x} and e_{d_y} such that either $c_i > d_x > c_j > d_y$ or $d_x > c_i > d_y > c_j$. Assume, without loss of generality, that $c_i > d_x > c_j > d_y$. A block interchange that acts on these four source edges creates four new cycles C', C'', D', and D'': C' is formed by the path that goes from e_{c_i} to e_{c_j} with a source edge that joins the first and last vertices of this path; C'' is formed by the path that goes from e_{c_j} to e_{c_i} with a source edge that joins the first and last vertices of this path; D' and D'' are analogous. The first segment of the block interchange starts at the source edge d_y, including the segment to be removed from e_{d_y} if the edge e_{d_y} is labeled, and ends at the source edge c_j. The second segment starts at the source edge d_x and ends at the source edge c_i, including the segment to be removed from e_{c_i} if the edge e_{c_i} is labeled. In this way, the segments to be removed from the same cycle are merged and the number of deletion runs remains the same, as shown in Fig. 6. $\qquad\square$

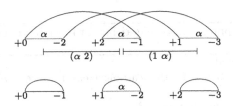

Fig. 5. Example of a block interchange acting on an oriented cycle and creating three new cycles. In this example, we have $A = (0\ \alpha\ 2\ \alpha\ 1\ \alpha\ 3)$ and $n = 2$. The block interchange moves the elements α in such a way that only one source edge remains labeled.

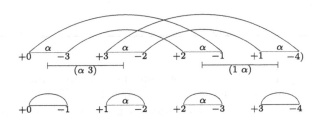

Fig. 6. Example of a block interchange acting on two non-oriented cycles and creating fours new cycles. In this example, we have $A = (0\ \alpha\ 3\ \alpha\ 2\ \alpha\ 1\ \alpha\ 4)$ and $n = 3$. The block interchange moves the elements α in such a way that the segments to be removed from the same cycle are merged.

Lemma 9. *For any strings A and ι^n, such that $|\pi^A| + 1 - c(A, \iota^n) = 0$, there exists a deletion ψ with $\Delta c(A, \iota^n, \psi) + \Delta\lambda(c, \iota^n, \psi) = 1$.*

Proof. Since $|\pi^A| + 1 - c(A, \iota^n) = 0$, each cycle of this graph is trivial. Each cycle has at most one insertion run and one deletion run. A deletion that cleans a source edge of a cycle C decreases the number of runs in C by one. Therefore, $\Delta\lambda(c, \iota^n, \psi) = 1$ and $\Delta c(A, \iota^n, \psi) = 0$. $\square$

Algorithm 1 uses the results of Lemmas 7 to 9.

Theorem 1. *Algorithm 1 is a 2-approximation for the problem of rearrangement distance with block interchanges and indels.*

Proof. By Lemmas 7 to 9, each operation $\beta \in \{\mathcal{BI}, \phi, \psi\}$ applied by the algorithm has $\Delta c(A, \iota^n, \beta) + \Delta\lambda(c, \iota^n, \beta) \geq 1$. In this way, at the end of the algorithm, the resulting string A' satisfies $|\pi^{A'}| + 1 - c(A', \iota^n) + \lambda(A', \iota^n) = 0$ and, consequently, $A' = \iota^n$. Furthermore, the algorithm uses at most $|\pi^A| + 1 - c(A, \iota^n) + \lambda(A, \iota^n)$ operations. By Lemma 5, the algorithm is a 2-approximation. $\square$

Algorithm 1: 2-Approximation algorithm for block interchange and indels distance.

    Input: Strings A and ι^n
    Output: An upper bound for the rearrangement distance $d_{\mathcal{BI}}(A, \iota^n)$
1 Let $d \leftarrow 0$
2 **while** $G(A, \iota^n)$ *has insertion runs* **do**
3 | Apply an insertion according to Lemma 7
4 | $d \leftarrow d + 1$
5 **while** $|\Sigma_A \cap \Sigma_{\iota^n}| + 1 - c(A, \iota^n) > 0$ **do**
6 | Apply a block interchange according to Lemma 8
7 | $d \leftarrow d + 1$
8 **while** $G(A, \iota^n)$ *has deletion runs* **do**
9 | Apply a deletion according to Lemma 9
10 | $d \leftarrow d + 1$
11 return d

For the BI and Reversal Distance, we consider gene orientation. Therefore, it is possible that divergent cycles exist in the labeled cycle graph. For convergent cycles, we can still apply only block-interchanges to create new cycles in the graph. The next lemma shows that it is always possible to find a reversal applied to a divergent cycle and break it into two new cycles, while maintaining the indel potential.

Lemma 10. *For any strings A and ι^n, such that $G(A, \iota^n)$ has no labeled target edges and $G(A, \iota^n)$ has a divergent cycle C, there exists a reversal ρ with $\Delta c(A, \iota^n, \rho) + \Delta\lambda(c, \iota^n, \rho) = 1$.*

Proof. Let $C = (c_1, c_2, \ldots, c_k)$ be a divergent cycle in $G(A, \iota^n)$ and let $(e_{c_x}, e_{c_{x+1}})$ be a pair of divergent edges with minimum x. A reversal applied to these edges breaks C into a trivial cycle C' and another cycle C'' [3]. The reversal can be chosen in a way that any α element is accumulated in the cycle C'', which makes the indel potential of the trivial cycle equals 0 and the indel potential of C'' equals the indel potential of C. In this way, we have $\Delta c(A, \iota^n, \rho) + \Delta\lambda(c, \iota^n, \rho) = 1$. An example of such operation is shown in Fig. 7. $\qquad\square$

Theorem 2. *Algorithm 2 is a 2-approximation for the problem of rearrangement distance with block interchanges, reversals, and indels.*

Proof. By Lemmas 7 to 10, each operation $\beta \in \{\mathcal{BI}, \rho, \phi, \psi\}$ applied by Algorithm 2 has $\Delta c(A, \iota^n, \beta) + \Delta\lambda(c, \iota^n, \beta) \geq 1$. After the while loop, the resulting string A' satisfies $|\pi^{A'}| + 1 - c(A', \iota^n) + \lambda(A', \iota^n) = 0$ and $A' = \iota^n$. Therefore, the algorithm uses at most $|\pi^A| + 1 - c(A, \iota^n) + \lambda(A, \iota^n)$ operations. By Lemma 6, the algorithm is a 2-approximation. $\qquad\square$

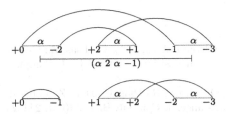

Fig. 7. Example of a reversal acting on a divergent cycle and creating two new cycles. In this example, we have $A = (0\ \alpha\ 2\ \alpha\ -1\ \alpha\ 3)$ and $n = 2$. The reversal moves the α element in such a way that the trivial cycle created has only clean edges.

Algorithm 2: 2-Approximation algorithm for Block Interchange, Reversal and Indel Distance.

Input: Strings A and ι^n
Output: An upper bound for the rearrangement distance $d_{\rho, \mathcal{BI}}(A, \iota^n)$
1 Let $d \leftarrow 0$
2 **while** $G(A, \iota^n)$ *has insertion runs* **do**
3 $\quad$ Apply an insertion according to Lemma 7
4 $\quad$ $d \leftarrow d + 1$
5 **while** $|\Sigma_A \cap \Sigma_{\iota^n}| + 1 - c(A, \iota^n) > 0$ **do**
6 $\quad$ **if** $G(A, \iota^n)$ *has a divergent cycle* **then**
7 $\quad\quad$ Apply a reversal according to Lemma 10
8 $\quad\quad$ $d \leftarrow d + 1$
9 $\quad$ **else**
10 $\quad\quad$ Apply a block interchange according to Lemma 8
11 $\quad\quad$ $d \leftarrow d + 1$
12 **while** $G(A, \iota^n)$ *has deletion runs* **do**
13 $\quad$ Apply a deletion according to Lemma 9
14 $\quad$ $d \leftarrow d + 1$
15 **return** d

The time complexity of both algorithms is $O(n^2)$. Creating the Labeled Cycle Graph and classifying its cycles takes $O(n)$ time. Each while loop runs for $O(n)$ iterations, and each operation can be performed in $O(n)$ time.

4 Conclusion

In this work, our main results are related to a structure called labeled cycle graph. This graph can represent a complete instance of the problems, and we were able to present good bounds for the Block Interchange and Indel Distance and the Block Interchange, Reversal, and Indel Distance. With these results, we developed 2-approximation algorithms for both distance problems.

The present study assumed equal costs for all rearrangements and absence of repeated genes in the genomes. To extend the research, future works can explore variations in the costs of rearrangements and the inclusion of genomes containing repeated genes.

References

1. Alexandrino, A.O., Oliveira, A.R., Dias, U., Dias, Z.: Genome rearrangement distance with reversals, transpositions, and indels. J. Comput. Biol. **28**(3), 235–247 (2021)
2. Alexandrino, A.O., Oliveira, A.R., Dias, U., Dias, Z.: Labeled cycle graph for transposition and indel distance. J. Comput. Biol. **29**(03), 243–256 (2022)
3. Bafna, V., Pevzner, P.A.: Genome rearrangements and sorting by reversals. SIAM J. Comput. **25**(2), 272–289 (1996)
4. Bafna, V., Pevzner, P.A.: Sorting by transpositions. SIAM J. Discret. Math. **11**(2), 224–240 (1998)
5. Braga, M.D., Willing, E., Stoye, J.: Double cut and join with insertions and deletions. J. Comput. Biol. **18**(9), 1167–1184 (2011)
6. Bulteau, L., Fertin, G., Rusu, I.: Sorting by transpositions is difficult. SIAM J. Discret. Math. **26**(3), 1148–1180 (2012)
7. Caprara, A.: Sorting permutations by reversals and eulerian cycle decompositions. SIAM J. Discret. Math. **12**(1), 91–110 (1999)
8. Chen, X.: On sorting permutations by double-cut-and-joins. In: Thai, M.T., Sahni, S. (eds.) COCOON 2010. LNCS, vol. 6196, pp. 439–448. Springer, Heidelberg (2010). https://doi.org/10.1007/978-3-642-14031-0_47
9. Christie, D.A.: Sorting permutations by block-interchanges. Inf. Process. Lett. **60**(4), 165–169 (1996)
10. El-Mabrouk, N.: Genome rearrangement by reversals and insertions/deletions of contiguous segments. In: Giancarlo, R., Sankoff, D. (eds.) CPM 2000. LNCS, vol. 1848, pp. 222–234. Springer, Heidelberg (2000). https://doi.org/10.1007/3-540-45123-4_20
11. Fertin, G., Labarre, A., Rusu, I., Tannier, É., Vialette, S.: Combinatorics of Genome Rearrangements. Computational Molecular Biology. The MIT Press, London (2009)
12. Hannenhalli, S., Pevzner, P.A.: Transforming cabbage into turnip: polynomial algorithm for sorting signed permutations by reversals. J. ACM **46**(1), 1–27 (1999)

13. Kececioglu, J.D., Sankoff, D.: Exact and approximation algorithms for sorting by reversals, with application to genome rearrangement. Algorithmica **13**, 180–210 (1995)

14. Willing, E., Stoye, J., Braga, M.: Computing the inversion-indel distance. IEEE/ACM Trans. Comput. Biol. Bioinf. **18**(6), 2314–2326 (2021)

15. Willing, E., Stoye, J., Braga, M.D.: Computing the inversion-indel distance. IEEE/ACM Trans. Comput. Biol. Bioinf. **18**(6), 2314–2326 (2020)

16. Yancopoulos, S., Attie, O., Friedberg, R.: Efficient sorting of genomic permutations by translocation, inversion and block interchange. Bioinformatics **21**(16), 3340–3346 (2005)

circTIS: A Weighted Degree String Kernel with Support Vector Machine Tool for Translation Initiation Sites Prediction in circRNA

Denilson Fagundes Barbosa[1,2], Liliane Santana Oliveira[1],
and André Yoshiaki Kashiwabara[1(✉)]

[1] Departamento Acadêmico de Computação (DACOM),
Programa de Pós-Graduação Associado em Bioinformática (PPGAB),
Universidade Tecnológica Federal do Paraná (UTFPR), Cornélio Procópio, Brazil
kashiwabara@utfpr.edu.br
[2] Instituto Federal de Educação, Ciência e Tecnologia de Santa Catarina (IFSC),
Canoinhas, Santa Catarina, Brazil

Abstract. Recent studies discovered that peptides generated from the translation of circRNAs participate in several biological processes, many related to human diseases. Researchers have observed that initiation of translation in circRNAs frequently occurs from non-AUG start codons. However, most existing computational tools for translation initiation site (TIS) prediction consider only the canonical AUG start codon. Thus, we developed a new methodology for predicting TIS AUG and near-cognates, considering the circularization of ORFs occurring in circRNAs. Initially, we used the weighted degree string kernel to create a data representation of the circRNA sequence fragments around possible TIS. Next, we applied a support vector machine to calculate a score representing the potential of the sequence fragment to contain an actual TIS. We used datasets from annotated TIS on circRNAs sequences to train and test our methodology. The first experiment showed that the sequence fragment length is the best value for the kernel's degree hyperparameter. Next, we investigated the most suitable sequence fragment length. Finally, we compared our methodology with three tools, TITER, TIS Predictor, and TIS Transformer. For TIS AUG prediction, circTIS obtained an AUROC of 98.64%, while TITER, TIS Predictor, and TIS Transformer obtained 78.97%, 78.39%, and 81.3%, respectively. For the TIS near-cognate prediction, our method obtained an AUROC equal to 96.84%, while TITER, TIS Predictor, and TIS Transformer got 81.37%, 72.68%, and 66.33%, respectively. We implemented our methodology in the circTIS tool, freely available at https://github.com/denilsonfbar/circTIS.

Keywords: circRNA · Translation initiation site prediction · Weighted degree string kernel · Support vector machine

1   Introduction

Circular RNAs (circRNAs) have a covalent link between their ends, making them more stable than linear RNA [13]. Cells usually produce circRNAs from known exons by a type of alternative splicing denoted as back-splicing [10]. With the development of next-generation sequencing, researchers have identified thousands of circRNAs in various organisms, with many having evolutionary conservation [9,12].

Despite the attention received in the last decades, the exact mechanism and function of most circRNAs still need to be discovered [23]. Recent studies have found evidence indicating that circRNAs have functions related to their translation [1,3,19]. However, new tools and experiments are required to understand their role, particularly in disease onset and progression [2,5,7,11,22].

Usually, tools for predicting coding regions in RNAs consider only AUG as a possible translation initiation site (TIS), ignoring non-AUG TIS that occurs in circRNAs [20]. We found three methodologies for predicting non-AUG TIS in linear RNAs, including PreTIS [16], TITER [25], and TIS Predictor [6]. PreTIS uses a linear regression model to predict TIS AUG and near-cognates. TITER employs deep learning to predict TIS and TIS Predictor uses Random Forest models to identify TIS in DNA sequence fragments associated with neurological disorders. These methodologies differ in the length of the input sequences, the training data, and the methods used to extract features. Nonetheless, all three return a score representing the probability of a sequence fragment containing an actual TIS.

Kernel methods are a powerful class of machine learning algorithms that allow the analysis and integration of different data types, using a kernel matrix representing the data and several algorithms to analyze it [18]. These methods use a kernel function that calculates a similarity measure to all pairwise combinations of data points, allowing data embedding in a suitable vector space and simplifying complex relationships between data.

String kernels are kernel functions that allow the comparison of strings. The weighted degree (WD) is a string kernel that compares strings by counting equal substrings in the same positions of original strings of length defined by the degree hyperparameter [15]. Researchers have widely used string kernels in Bioinformatics for sequence analyses [14].

Support vector machine (SVM) is the most widely used kernel method, which maps the data into a higher-dimensional feature space and constructs a separating hyperplane with the broadest possible margin to perform binary classification between classes [17]. SVM also allows the distance of a data point from the separating hyperplane to quantify the magnitude of the point belonging to a given class.

Because of the lack of specific software for TIS prediction in circRNA, we developed circTIS, a tool based on string kernel and support vector machine, detailed in the following sections.

2   Materials and Methods

This section details how the circRNAs were selected and split into test and training datasets. Next, we present our proposed methodology for TIS prediction, the experimental setup for selecting the WD kernel degree and C hyperparameters, and the length of sequence fragments. Lastly, we explain the method for comparison with existing tools for predicting TIS AUG and near-cognates.

2.1   circRNAs Selection and Datasets Construction

TransCirc [8] is a database that gathers 328,080 exonic circRNAs detected in human tissues. From an extensive literature search, the authors of TransCirc matched seven different types of translation evidence to circRNAs. Thus, a circRNA in TransCirc has none or different combinations of the seven translational evidence associated with it. The main types of translation evidence cataloged by TransCirc include data from mass spectrometry with encoded peptides across back-splice junctions, data from ribosome/polysome experiments that detected occupancy of ribosomes by circRNAs, and known TIS mapped on circRNA sequences.

For constructing our datasets, we initially selected the 9,394 circRNAs available in TransCirc with evidence of translation supported by one or more TIS, resulting in 10,636 associated TIS. To determine TIS positions in circRNA sequences, TransCirc authors mapped previously annotated TIS in circRNAs, from their genomic positions, available in TISdb [24]. TISdb is a database containing canonical and alternative TIS detected in human genes using the global translation initiation sequencing (GTI-Seq) technique. The TransCirc database stores many circRNAs with different identifiers but with the same nucleotide sequence and the same TISs mapped to these sequences. After we removed these repeats, 6,650 circRNAs with 7,665 annotated TIS remained, which made up our primary dataset.

Then, as seen in Table 1, we divided our primary dataset into six parts with approximately the same amount of circRNAs, separating one of these parts to

Table 1. Separation of the primary dataset into six parts with approximately the same amount of circRNAs. We also present the number of annotated TIS and the number of equivalent codons non-TIS, considered in our methodology, extracted from each set of circRNAs.

	Split 1	Split 2	Split 3	Split 4	Split 5	Test dataset
circRNAs	1,109	1,109	1,108	1,108	1,108	1,108
TIS AUG	878	861	869	820	858	881
non-TIS AUG	17,090	17,447	19,082	18,735	18,186	18,422
TIS near-cognates	116	112	104	104	96	98
non-TIS near-cognates	49,070	49,986	53,389	52,076	52,440	52,782

compose our test dataset. We use the test dataset for the final evaluation of the model and comparison with other similar tools. The other five parts composed our training dataset, used with 5-fold cross-validation. Ultimately, we grouped these five parts for training the final model implemented in circTIS.

From each circRNA, we extract fragments of the nucleotide sequence around possible TIS. Like the TITER and TIS Predictor authors, we consider the AUG codons as canonical TIS and the CUG, GUG, and UUG codons as near-cognate TIS. We disregarded the other annotated TIS, composed of codons with nucleotide combinations different from the four mentioned. We obtained the positions of the real TIS from TransCirc and considered as false TIS all existing codons in the circRNA sequences, regardless of the frame, corresponding to one of the four cited types. Table 1 shows the true and false TIS extracted from each circRNA set.

2.2 Proposed Methodology for TIS Prediction

Our methodology consists of creating a representation of the samples constituted by sequence fragments around possible TIS using the WD kernel and training an SVM model to classify the samples. Classification is performed based on a score corresponding to the distance of the sample concerning the separator hyperplane of the trained SVM. As sequence circularization occurs in circRNAS, all our samples have the same length in each experimental configuration, which is necessary for using the WD kernel.

Before the SVM's definitive training, we conducted experiments to select the best values for the WD kernel's hyperparameter degree and the SVM's hyperparameter C. We also performed previous experiments to determine the best fragment size around possible TIS to perform their identification in circRNAs. We carried out these previous experiments using the 5-fold cross-validation technique, in which, for each configuration tested, we trained the model with four parts of the training dataset and used the fifth part for validation, repeating this procedure five times using each a different part of the training dataset for validation. Before performing each training step, we randomly downsampled the negative samples, matching their number to the positive ones.

Our first experiment aimed to investigate the best value for the degree hyperparameter of the WD kernel. The WD kernel compares two strings by counting all occurrences where two substrings, one from each input string, are equal. WD is a positional kernel, as the substrings occupy the same positions in the input strings. The degree defines the maximum length of substrings for comparison. This experiment used samples with 203 nucleotides (100 downstream and 100 upstream of the possible TIS). Then, setting the degree value to one, we train the SVM and evaluate the classification performance. We then repeated the training and testing by varying the degree hyperparameter from two to 203, given that the degree value limit is the sequence sample size. As presented in the Results section, the classification quality increased as we increased the degree value. Thus, in the following experiments, we set the WD kernel degree value to the length of the sequence fragments we used as samples.

The following experiments aimed to determine the length of the sequences that yield the best classification performance. Here, we used samples with 403 nucleotides (100 downstream and 300 upstream of the possible TIS). Initially, we reduced the sequences of training and validation to just one nucleotide in the downstream direction. Then, using these reduced samples, we executed cross-validation. We then repeated the training and testing by increasing the sample in the downstream sense one nucleotide at a time until we reached the limit of 303 nucleotides. We then evaluated which sequence length generated the best classification.

Similarly, we again left the sample with one nucleotide. Now, we were incrementing and evaluating the classification result, increasing the sample upstream up to the limit of 100 nucleotides. Again, we evaluated which sample length generated the best classification. The selected length of the samples we used in the circTIS final model corresponds to the best upstream sequence length concatenated to the best downstream sequence length founds, which we present in the Results section.

We also realized experiments for selecting the best value for the C hyperparameter. The C hyperparameter balances the maximization of the margin of the SVM hyperplane with the minimization of training errors. Initially, we tested the values $2^{-3}, 2^{-2}, ..., 2^6$, obtaining the best result with the value 2. Then, we performed fine-tuning by testing the values $0.5, 0.75, ..., 2.5$, where we got the best result with the value 1.5. Finally, we tested the values in the interval $1.25, 1.3, ..., 1.75$, with the best result at 1.55. Therefore, we set the hyperparameter C to 1.55 in the final model.

We used the metrics F1-score, precision, and recall to evaluate the predictions of these training experiments. As the validation dataset is significantly unbalanced, the F1-score is more suitable for evaluating these results. The F1-score is given by the harmonic mean between precision and recall, representing the predictor's ability to correctly identify the samples with TIS simultaneously regarding the number of samples incorrectly classified as containing or not a TIS.

For the software implementation, we used Python scripts with the public Shogun machine learning library version 6.1.3 [21] to compute kernel matrices, construction, and manipulation of the SVM. We performed the experiments on a personal computer with an operating system Debian-based with Linux kernel version 5.15.0-67. The hardware comprises an 11th-generation Intel i7 processor and 16 GB of main memory.

2.3   Comparison with Existing Methodologies

To evaluate the circTIS, we compared the performance of our model with TITER and TIS Predictor using our test dataset. We also compare our method with TIS Transformer [4], a recently published deep learning-based tool specialized for TIS AUG prediction. We did not retrain the models in these experiments, so the analysis of the results must consider that the authors originally trained them with linear RNAs. Until the writing of this article, we did not find specific tools for predicting TIS in circular RNAs to compare our methodology.

In this comparison, we submitted our test dataset for classification by the three cited tools. For the TITER, we used samples with 203 nucleotides (100 downstream and 100 upstream of the possible TIS, according to the sample length defined by your authors). We prepared samples with only 23 nucleotides (ten downstream and ten upstream of the possible TIS) for TIS Predictor, as described in their release article. These sample lengths cannot be changed as they are part of the architecture of the respective tools. TIS Transformer accepts complete RNA sequences as inputs, so we submit the FASTA file of the test dataset with complete circRNAs, without prior extraction of samples of sequence fragments. We did not make any further changes in models or parameters to perform the comparisons.

As the methodologies compared return a score measuring the potential of a sample (or of each sequence position, in the case of the TIS Transformer), we analyzed the results visually from graphs containing the ROC (Receiver Operating Characteristic) and PR (Precision-Recall) curves generated from the samples extracted from the test dataset. We also used the values of the areas under the curves for a numerical comparison of the results.

3   Results

Next, we show the results obtained in the experiments for selecting the WD kernel degree, the length of the sequence fragment, and the comparison with similar programs.

3.1   WD Kernel Degree Selection

In this experiment, we investigated the best value for the WD kernel degree hyperparameter. We used samples with 203 nucleotides (100 downstream and 100 upstream of the possible TIS). We consider the value of the F1-score to carry out the final evaluation.

As can be seen in Fig. 1, we found that, for the datasets used, there was a constant improvement of the F1-score as we increased the degree value. The limit for the degree value is the size of the sequences we use as samples. Thus, we selected a degree value equal to the sequence length for training the model in the following experiments. In the biological context, this result suggests that the WD kernel could use the full extent of the samples to classify them as containing or not a TIS.

3.2   Length of the Sequence Sample Selection

Now, we show the experiment results varying the sample size, with different window lengths downstream and upstream of the possible TIS, indicating which window sizes produce the best results for the problem under study.

Considering the F1-score in Fig. 2 (left), we can observe that the region upstream of the possible TIS has a minor influence on the correct classification of the samples. From a length above approximately 40 positions, we can see

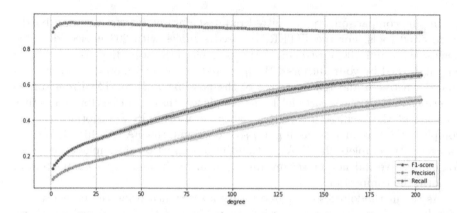

Fig. 1. Result of the experiment to investigate the best value for the WD kernel degree hyperparameter. Considering the F1-score, it is possible to observe a constant improvement in the result as we increase the degree value. The dots in the graph are the average values from the 5-fold cross-validation.

a soft and constant decrease in the F1-score. Thus, we selected the length of 40 nucleotides upstream of TIS to build the final model implemented in circTIS.

In Fig. 2 (right), we can see an improvement in the classification, even around a sample with more than 250 nucleotides in length. From this position, the classification quality presents an establishment. The best F1-score occurs in this experiment with a sample of 287 nucleotides. Therefore, we selected this length downstream TIS to build the final model implemented in circTIS.

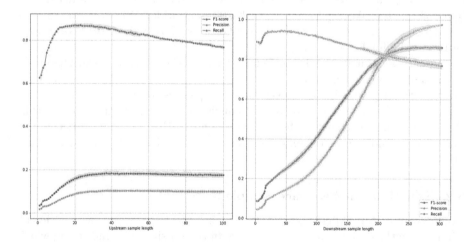

Fig. 2. Classification quality by varying the sample size upstream (left) and downstream (right) to the possible TIS. The dots in the graphs are the average values from the 5-fold cross-validation.

Therefore, based on these results, we defined a length of 327 positions for extracting fragments of circRNA sequences manipulated by circTIS, 40 upstream and 287 in the downstream direction from the possible TIS to be classified.

3.3   Comparison with Similar Tools

We performed this comparison by evaluating the ability of the models to predict TIS AUG and TIS near-cognates (CUG, GUG, and UUG, in this study) separately. As shown in Fig. 3, TITER, TIS Predictor, and TIS Transformer generated similar results. However, the circTIS significantly outperformed these tools. Considering the ROC curve of TIS AUG prediction, circTIS obtained an AUROC of 98.64%, while TITER, TIS Predictor, and TIS Transformer obtained 78.97%, 78.39%, and 81.3%, respectively. The AUPR value of our method was 92.42%, while TITER, TIS Predictor, and TIS Transformer reached 18.82%, 20.50%, and 30.21%, respectively.

In the second comparison at the bottom of Fig. 3, we substituted the TIS Predictor model for TIS near-cognate prediction according to the TIS type since its authors trained a specific model to predict TIS AUG and another for near-cognate TIS. Again, our methodology significantly outperformed the other tools. Our method obtained an AUROC equal to 96.84%, while TITER, TIS Predictor, and TIS Transformer got 81.37%, 72.68%, and 66.33%, respectively. The AUPR value of our method was equal to 74.22%, while TITER, TIS Predictor, and TIS Transformer reached 4.25%, 0.89%, and 0.64%, respectively.

Regarding the processing time, circTIS took 94 s to calculate the kernel matrix and train the SVM final model with the complete balanced training dataset. To perform the prediction of samples in the test dataset, circTIS needed 382 s, with most of this time spent extracting samples from circRNA sequences. To perform the prediction on the same test dataset, TITER took 2,295 s, TIS Predictor needed 2,078 s, and TIS Transformer needed only 178 s. The hardware comprises an 11th-generation Intel i7 processor and 16 GB of main memory.

4   Discussion

This paper introduces a new computational method for predicting TIS AUG and near-cognates in circRNA sequences. The method involves extracting samples composed of fragments of sequences around potential TIS and classifying them as having or not an actual TIS.

Our method is similar to most TIS prediction tools for linear circRNA. However, we employ a different approach to prediction, using the WD string kernel to represent samples and SVM to perform classification. Since circularization of the ORFs occurs in circRNAs, we found one advantage compared to other tools developed for linear RNAs: we did not use samples with partial sequences. All samples used in our experiments were sequences of the same length.

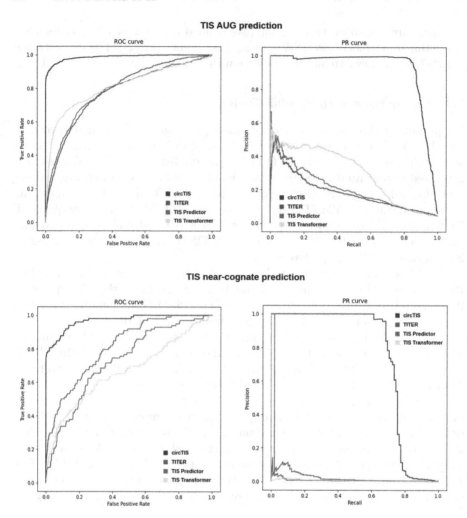

Fig. 3. ROC and PR curves generated by the four methods for the TIS AUG and TIS near-cognate classification. TITER, TIS Predictor, and TIS Transformer generated similar results. The circTIS significantly outperformed the other three tools.

The first contribution of this work was the construction and release of datasets with TIS AUG and near-cognates of circRNA sequences. These datasets and all scripts used for experimentation are available at https://github.com/denilsonfbar/circTIS-exps-BSB2023.

Our primary contribution, which we did not find in the literature, was the observation that the WD kernel hyperparameter degree equal to the size of the input sequences produces better results for TIS classification in circRNAs. In future work, we will investigate this finding through experiments on other datasets and for other sequence classification problems, such as predicting TIS in linear RNAs.

Another significant contribution was the investigation of the best lengths downstream and upstream of the potential TIS for classification. In the context of circRNA fragments represented with the WD kernel, we found a limit of around 250 nucleotides downstream for the best discrimination between real and false TIS. As expected, this length is smaller in the upstream direction, around 40 nucleotides. While more experiments are needed, this result suggests that tools like TITER, which uses samples with 100 nucleotides upstream, could be more efficient in reducing samples in that direction.

Finally, another contribution is the implementation of the proposed methodology in a tool called circTIS and its release for use by other researchers. circTIS also stands out for its fast training and prediction times. circTIS is freely available at https://github.com/denilsonfbar/circTIS.

5   Conclusion

The initial experiments we presented suggest that the circTIS constitutes a valuable tool for predicting TIS in circRNAs. However, new experiments are needed to confirm the initial results presented. Although circTIS showed significantly better results than the other tools, it is essential to highlight that these other models were trained for linear RNAs. Although the circRNAs used in our experiments result from alternative splicing of known exons, the biological particularities of the circRNAs, used for circTIS training require that new comparisons be performed to confirm the presented results.

References

1. Abe, N., et al.: Rolling circle translation of circular RNA in living human cells. Sci. Rep. **5**, 1–9 (2015). https://doi.org/10.1038/srep16435
2. Aufiero, S., Reckman, Y.J., Pinto, Y.M., Creemers, E.E.: Circular RNAs open a new chapter in cardiovascular biology. Nat. Rev. Cardiol. **16**(8), 503–514 (2019). https://doi.org/10.1038/s41569-019-0185-2
3. Chen, C.Y., Sarnow, P.: Initiation of protein synthesis by the eukaryotic translational apparatus on circular RNAs. Science **268**(5209), 415–417 (1995). https://doi.org/10.1126/science.7536344. www.science.org/doi/10.1126/science.7536344
4. Clauwaert, J., McVey, Z., Gupta, R., Menschaert, G.: TIS transformer: remapping the human proteome using deep learning. NAR Genom. Bioinform. **5**(1), 1–8 (2023). https://doi.org/10.1093/nargab/lqad021
5. Fang, Y., et al.: Screening of circular RNAs and validation of circANKRD36 associated with inflammation in patients with type 2 diabetes mellitus. Int. J. Mol. Med. **42**(4), 1865–1874 (2018). https://doi.org/10.3892/ijmm.2018.3783
6. Gleason, A.C., Ghadge, G., Chen, J., Sonobe, Y., Roos, R.P.: Machine learning predicts translation initiation sites in neurologic diseases with nucleotide repeat expansions. PLoS ONE **17**(6 June), 1–30 (2022). https://doi.org/10.1371/journal.pone.0256411. www.dx.doi.org/10.1371/journal.pone.0256411
7. Hanan, M., Soreq, H., Kadener, S.: CircRNAs in the brain. RNA Biol. **14**(8), 1028–1034 (2017). https://doi.org/10.1080/15476286.2016.1255398

8. Huang, W., et al.: TransCirc: an interactive database for translatable circular RNAs based on multi-omics evidence. Nucleic Acids Res. **49**(D1), D236–D242 (2021). https://doi.org/10.1093/nar/gkaa823

9. Jeck, W.R., et al.: Circular RNAs are abundant, conserved, and associated with ALU repeats. RNA **19**(2), 141–157 (2013). https://doi.org/10.1261/rna.035667. 112

10. Kristensen, L.S., Andersen, M.S., Stagsted, L.V., Ebbesen, K.K., Hansen, T.B., Kjems, J.: The biogenesis, biology and characterization of circular RNAs. Nat. Rev. Genet. **20**(11), 675–691 (2019). https://doi.org/10.1038/s41576-019-0158-7

11. Li, H., et al.: Comprehensive circular RNA profiles in plasma reveals that circular RNAs can be used as novel biomarkers for systemic lupus erythematosus. Clinica Chimica Acta **480**(Jan), 17–25 (2018). https://doi.org/10.1016/j.cca.2018.01.026

12. Memczak, S., et al.: Circular RNAs are a large class of animal RNAs with regulatory potency. Nature **495**(7441), 333–338 (2013). https://doi.org/10.1038/nature11928

13. Patop, I.L., Wüst, S., Kadener, S.: Past, present, and future of circRNAs. EMBO J. **38**(16), 1–13 (2019). https://doi.org/10.15252/embj.2018100836

14. Qi, R., Guo, F., Zou, Q.: String kernels construction and fusion: a survey with bioinformatics application. Front. Comput. Sci. **16**(6), 166904 (2022). https://doi.org/10.1007/s11704-021-1118-x

15. Ratsch, G., Sonnenburg, S.: Accurate splice site detection for Caenorhabditis Elegans. In: Kernel Methods in Computational Biology. The MIT Press (2004). https://doi.org/10.7551/mitpress/4057.003.0018

16. Reuter, K., Biehl, A., Koch, L., Helms, V.: PreTIS: a tool to predict non-canonical 5' UTR translational initiation sites in human and mouse. PLoS Comput. Biol. **12**(10), 1–22 (2016). https://doi.org/10.1371/journal.pcbi.1005170

17. Schölkopf, B., Smola, A.J.: Learning with Kernels. The MIT Press, Cambridge (2018). https://doi.org/10.7551/mitpress/4175.001.0001. www.direct.mit.edu/books/book/1821/learning-with-kernelssupport-vector-machines

18. Shawe-Taylor, J., Cristianini, N.: Kernel Methods for Pattern Analysis. Cambridge University Press, Cambridge (2004). https://doi.org/10.1017/CBO9780511809682. www.cambridge.org/core/product/identifier/9780511809682/type/book

19. Shi, Y., Jia, X., Xu, J.: The new function of circRNA: translation. Clin. Transl. Oncol. **22**(12), 2162–2169 (2020). https://doi.org/10.1007/s12094-020-02371-1

20. Sinha, T., Panigrahi, C., Das, D., Chandra Panda, A.: Circular RNA translation, a path to hidden proteome. Wiley Interdiscip. Rev. RNA **13**(1), 1–15 (2021). https://doi.org/10.1002/wrna.1685

21. Sonnenburg, S., et al.: The Shogun machine learning toolbox. J. Mach. Learn. Res. **11**(June), 1799–1802 (2010)

22. Vo, J.N., et al.: The landscape of circular RNA in cancer. Cell **176**(4), 869–881.e13 (2019). https://doi.org/10.1016/j.cell.2018.12.021. www.linkinghub.elsevier.com/retrieve/pii/S0092867418316350

23. Vromman, M., Vandesompele, J., Volders, P.J.: Closing the circle: current state and perspectives of circular RNA databases. Brief. Bioinform. **22**(1), 288–297 (2021). https://doi.org/10.1093/bib/bbz175

24. Wan, J., Qian, S.B.: TISdb: a database for alternative translation initiation in mammalian cells. Nucleic Acids Res. **42**(D1), 845–850 (2014). https://doi.org/10.1093/nar/gkt1085

25. Zhang, S., Hu, H., Jiang, T., Zhang, L., Zeng, J.: TITER: predicting translation initiation sites by deep learning. Bioinformatics **33**(14), i234–i242 (2017). https://doi.org/10.1093/bioinformatics/btx247

Evaluating the Molecular—Electronic Structure and the Antiviral Effect of Functionalized Heparin on Graphene Oxide Through *Ab Initio* Computer Simulations and Molecular Docking

André Flores dos Santos(✉) , Mirkos Ortiz Martins , Mariana Zancan Tonel ,
and Solange Binotto Fagan

Postgraduate Program in Nanosciences: Laboratory of Simulation and Modeling of
Nanomaterials-LASIMON, Franciscan University-UFN, Santa Maria, RS, Brazil
`andre.santos@ufn.edu.br`

Abstract. In antiviral studies, heparin is widely used against the SARS-CoV-2 virus. In this study, computer simulations were performed to understand the role of heparin in a possible blockade of the spike protein binding with the human cell receptor. Another molecule, graphene oxide (GO), was functionalized to interact and bind with heparin to achieve an increase in binding affinity with the spike protein. In the first stage. The electronic and chemical interaction between the molecules were analyzed through *ab initio* simulations by using Spanish Initiative for SIESTA (Electronic Simulations with Thousands of Atoms) Software. Next, we evaluated the interaction between molecules together and separately in the spike protein target through molecular docking simulations using *AutoDock Vina* Software. The results were relevant because GO functionalized with heparin exhibited an increase in affinity energy to the spike protein. This affinity indicated a possible increase in antiviral activity. This increase will be verified in the future through in vitro tests. Experimental tests on the synthesis and morphology of the material preliminarily indicate a good interaction between molecules and absorption of heparin by GO. This phenomenon confirmed the results of first principles simulations.

Keywords: coronavirus · drugs · pandemic

1 Introduction

Since the COVID-19 pandemic began in 2020, almost 6,881,955 (as of April 2023) million deaths have been reported worldwide [11]. Several treatment methods against the disease are based on experiments with similar viruses, such as severe acute respiratory syndrome coronavirus (SARS-CoV), acquired immunodeficiency syndrome (HIV), middle east respiratory syndrome (MERS-CoV), and virus influenza (H1N1). Antivirals explore a specific pathway in the COVID-19 infection process and are classified as endosomal acidification inhibitors, membrane fusion inhibitors, protein and viral entry blockers, viral replication blockers, and protease inhibitors [30].

© The Author(s), under exclusive license to Springer Nature Switzerland AG 2023
M. S. Reis and R. C. de Melo-Minardi (Eds.): BSB 2023, LNBI 13954, pp. 25–35, 2023.
https://doi.org/10.1007/978-3-031-42715-2_3

In previous studies, we analyzed antiviral drugs with the targets involved in the infectious process of COVID-19 without the use of nanostructures [21]. In this study, the spike protein was the target, and heparin separated and linked to graphene oxide (GO) was the ligand. This study verified the blocking activity of the chosen molecules in the binding process of the spike protein with the cell receptor (ACE2). In our research group, a similar theme using GO and flavonoids was used to propose a methodology to the current method, but with the molecules separated [22].

Heparin, widely used for the treatment of COVID-19 related to thrombosis and clots, is being investigated as a viral entry blocker. Some studies have proved the effectiveness of this drug. For example, a study showed that the dependence of heparin in the connection between the viral protein and cellular receptor and indicated that heparin can be used as a blocker of the SARS-CoV-2 spike protein [3]. Furthermore, heparin can be used against virus transmission [10]. Heparin inhibits cellular invasion by the SARS-CoV-2 virus by up to 80% [15]. A study proved the efficacy of SARS-CoV-2 inhibition by heparin and its sulfated polysaccharide derivatives and demonstrate its antiviral efficacy and its potential for prophylactic and therapeutic purposes [26]. A study investigated heparin as a cofactor for angiotensin-converting enzyme 2 (ACE2) in binding with the spike protein [33].

GO has been used in some virus research because of its excellent binding characteristics (negative charges), large surface area, and ease of functionalization with targeted drugs [9, 20, 24]. GO has been used for the treatment of other diseases such as cancer [25].

Studies have focused on the use of heparin with GO alone or associated with other molecules against COVID-19. For example, a study on the antimicrobial activity of the dialdehyde cellulose-graphene oxide film indicated the antimicrobial and antiviral activity of GO [4]. In another study, an in-depth investigation to understand the inactivation route of SARS-CoV-2 using GO revealed excellent results for virus inactivation when in contact with the carbon nanostructure [6]. In similar investigation, GO interferes with the surface of SARS-CoV-2 by exploring the sensitive regions of a variant of the spike protein and the angiotensin-converting enzyme of the cellular viral entry receptor 2 (ACE2) [28]. Molecular dynamics simulations on the interactions between SARS-CoV-2 and 3CL-Mpro by graphene and its derivatives revealed excellent absorption and inactivation of the viral protein [31]. Studies on nanotoxicity, which can cause serious problems for humans, have used graphene and GO in biological media and require meticulous attention so that cell death is not altered during experiments [23]. Another study investigated how to treat GO to obtain biocompatibility with human cells and deliver drugs to their target without problems [16]. A study detailed advances in graphene-based materials in drug delivery applications, including antitumor drugs, photodynamic therapy applications and optical imaging in theranostics, and exhibited excellent results and a promising future [12].

In this study we analyzed the improvement in the blocking activity associated with the binding of the spike protein with the human cellular receptor when GO was combined with heparin. As previously reported, the two molecules exhibited antiviral characteristics. However, the results of the two molecules together are yet to be investigated. According to the results of our molecular docking test, an increase in the affinity energy was observed between the bound molecules and the spike protein. This increase possibly indicates an improvement in the blocking activity of the spike protein and viral infection. *In vitro* and *in vivo* tests have been conducted to compare the results of the simulations and verify the cytotoxicity that the molecules can cause in cells and identify an adequate amount of dosage to be administered in humans against the SARS-CoV-2 virus as well as other viruses of similar families and with characteristics.

2 Methodology

Figure 1 shows the methodology used in this work, with the appointment that there are steps that in progress.

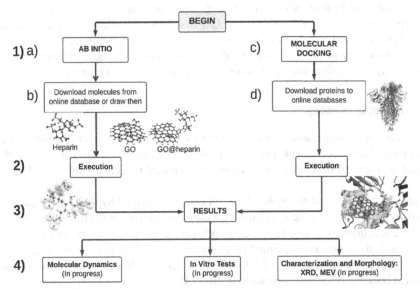

Fig. 1. 1-a: Ab initio methodology, b: download of the heparin molecule from the Pubchem Database, and creation of the GO in the Chemcraft Software, c: Molecular Docking methodology, d: Download of proteins from the Protein Data Bank (PDB). 2- Execution of the simulations. 3- Results analysis. 4- Steps in progress: Molecular dynamics, in vitro tests and characterization and morphology tests (DRX and SEM)

2.1 Ab Initio

The computational simulation of first principles, or *ab initio*, is used to analyze and predict the interactions of molecular systems and study their electronic and structural properties as well as interpret experimental results [18]. In *ab initio* simulations, the results were obtained through the hypotheses and basic equations of quantum mechanics, with charge, mass, and fundamental constants as parameters.

Ab initio simulations were performed using the SIESTA (Spanish Initiative for Electronic Simulations with Thousands of Atoms) Software, in which efficient electronic structure calculations, such as density functional theory, are implemented [7, 13]. The calculations were set to converge when the residual force on each atom was less than 0.05 eV/Å. The supercell was configured with a size of 45 Å in the X-direction, and 30 Å in the Y- and Z-directions.

The binding energy (EB) is given by Eq. (1), where Eb (GO + Heparin) corresponds to the total energy of the interaction system when graphene oxide interacts with heparin. Here E_{GO} and $E_{Heparin}$ are the total energy corresponding to each isolated molecule, heparin, and GO.

$$E_b = \left\{ E_{(Go+Heparin)} - \left(E_{GO} + E_{Heparin} \right) \right\} \tag{1}$$

GO and heparin molecules were initially tested separately to describe their electronic properties. Two configurations of GO and heparin molecules were obtained: GO_single_face with the molecular formula ($C_{55}H_{21}O_6$) is a configuration with functional groups only at the top, and GO_double_face GO with the molecular formula ($C_{56}H_{24}O_9$) is a configuration with functional groups on top and bottom.

2.2 Molecular Docking

In molecular modeling, molecular docking is a mechanism that predicts the preferential orientation of a macromolecule (usually a protein, peptides, or a stretch of DNA) called a target, to a second structure called a ligand when joined to form a stable complex [21, 27].

The fitting process at the molecular level can be described by the laws of quantum chemistry in which the time evolution of molecular systems is expressed in terms of the wave functions of atoms. However, in practice, approximations are used such that the system's dynamics are identified by atoms represented by point masses that move in the fields of molecular forces. Molecular forces are established by electrostatic and chemical bonding interactions between atoms [14].

Molecular fit is a placement relationship of a ligand to a target molecule and binding score based on some metrics, such as, the scoring function and root mean square deviation (RMSD), defined as a measure of the distance average in Angstrom between the atoms of the two ligands (receptor and ligand). RMSD is used to measure the quality of the molecular docking process, which indicates a value of fewer than two Angstroms preferably, whereas the affinity value should be as negative as possible [27].

AutoDock Vina integrated with AMDockTools [29] was used to perform molecular docking. The initial steps of docking, namely ligand preparation and receptor analysis, were performed using AutoDockTools Software that is part of the AutoDock Suite. The target molecule, also called the receptor, should be treated with the identification of charges, correction of unbound atoms for the stabilization of the structure, and solvation of the medium with water molecules [5]. Torsion was added to the GO@Heparin ligand, and its structure was analyzed to identify possible torsion sites to allow its adaptation to various spatial conformations during the execution of the docking. The results revealed angular movements in its three-dimensional structure.

The structures used in this study were obtained from an online database. The heparin molecule ($C_{12}NH_{15}S_3O_{19}$) was obtained from the PubChem molecules database with ID (5288499) [19]. Single-face ($C_{55}H_{21}O_6$) and double-face ($C_{56}H_{24}O_9$) molecules were produced at our nanomaterials simulation laboratory using *ChemCraft Software* [2]. The structures of the spike protein of SARS-CoV2 were obtained from the PDB Protein Data bank with ID (PDB ID: 7XNQ) [17]. The coordinates for the spike protein (7XNQ) binding domain (RBD) region with the cell receptor were mapped with the following data (X:146.9; Y:151.6; Z:117.5; SizeX:57.0; SizeY:54; SizeZ:52). Notably, only this region of the protein was evaluated because it is the connection between viral entry and host cells [21]. The structure of the spike protein (7XNQ) is displayed in Fig. 2 with the receptor-bind domain (RDB).

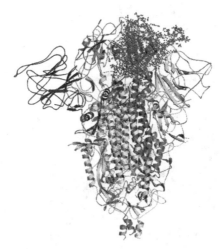

Fig. 2. The structure of the Spike Protein Omicron Variant BA.4 (7XNQ), with the Receptor-Bind Domain (RBD) region highlighted in pink

3   Results and Discussion

A*b initio* computer simulations results achieved excellent attraction responses between GO and heparin molecules in the regions where we approximated between atoms. The total energy and binding energy were the parameters for the four configurations shown in Table 1.

Table 1 describes the binding energy, the difference between the energy bands (HOMO and LUMO), charge transfer, the types of bonds made, and the distances between atoms. The configuration with the major binding energy was version 4, where two regions of heparin attraction with GO were highlighted. The first region was the one that made a hydrogen bond with GO. The other region of attraction of heparin with GO was a sulfated zone, but without binding.

Figure 3 displays the image of the four versions of heparin being approximated to GO in various regions, and their characteristics are presented in Table 1. In version 1, a

Table 1. Results of the *ab initio* computational simulations of some evaluation parameters.

Conformation	Ebind (eV)	H-L (eV)	Charge Transfer (e-)	Type of bond	Distance (Å-Angstrom)
Version 01	−2.740	0.064	0.649	H-O	1.6
Version 02	−2.000	0.060	0.709	C-O	1.6
Version 03	−1.980	0.053	0.801	H-O	1.2
Version 04	−3.460	0.028	0.968	H-O	1.2

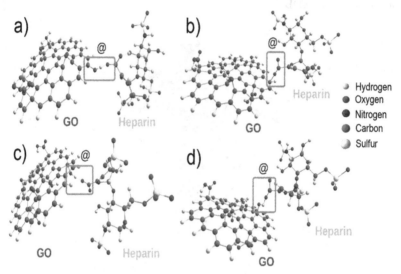

Fig. 3. Graphene oxide and heparin molecules approximated in lateral regions and functional groups a) version 1, b) version 2, c) version 3, and d) version 4

heparin oxygen was approximated to a hydrogen of a lateral hydroxyl of GO. In version 2, one oxygen of heparin was approximated to one carbon of one hexane of GO. In version 3, a heparin oxygen was approximated to a hydroxyl at the bottom of the GO. In version 4, a heparin oxygen was approximated to a hydroxyl in the upper part of the GO. The basic idea was to test different regions of approximation between the molecules, in order to verify the interaction between them.

Figure 4 shows the details of the energy bands (HOMO and LUMO) for double-sided GO, heparin alone and GO@heparin version 4. It is possible to observe that heparin is more electronegative than GO, and after the union of the molecules the energy bands were at an intermediate value compared to the molecules alone, which is an expected behavior.

Molecular docking results revealed a good result when GO was used with heparin, and the binding affinity increased. In this scenario, the antiviral blockade may increase in practical tests.

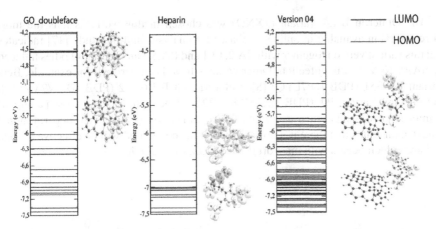

Fig. 4. Description of energy bands for graphene oxide molecules, heparin isolated and GO@Heparin (interacting molecules)

Various types of torsions, namely torsions 4, 8, and 16, were investigated for the GO@Heparin ligand. The maximum number admitted for GO@Heparin was 16 because when the number was greater than this number, the RMSD presented values above 2 Angstroms, which were considered a low-quality connection [1, 27]. Therefore, tests should be performed on the most rigid and least flexible molecule.

Table 2. Molecular Docking results with the Spike Protein 7XNQ (Omicron Variant BA.4)

Conformation	Torsion 4 (Affinity kcal/mol)	RMSD (Å)	Torsion 8 (Affinity kcal/mol)	RMSD (Å)	Torsion 16 (Affinity kcal/mol)	RMSD (Å)
Version 1	−16.9	0.214	−14.7	0.709	−12.8	1.609
Version 2	−15.4	0.281	−13.9	0.294	−11.4	1.060
Version 3	−14.4	0.283	−12.8	0.293	−10.9	1.090
Version 4	−13.8	0.171	−12.8	0.300	−10.8	0.888
Heparin	−8.6	0.360	−8.1	1.170	−7.5	1.580

Table 2 describes the conformations used in this study, the number of torsions used in ligands 4, 8, and 16, and their respective affinity values (kcal/mol), RMSD for spike protein 7XNQ. As expected, GO@Heparin always achieved superior values compared to heparin isolated. Furthermore, when the value of the number of torsions of the ligand increased, the RMSD value also increased and the affinity value decreased linearly for all values.

For the calculation of RMSD, the original ligand (with defined torsion) was considered in comparison with the first configuration of the results, that is, with a higher affinity value.

The Omicron BA.4 Variant (7XNQ) was chosen in this work, as it is the most current version available in online databases for analysis, and the study [17] indicates that this variant version together with BA.2.12.1 and BA.5, can evade antibodies induced by SARS-CoV-2 virus infection. Other variants have been verified, for example, Beta Variant B.1.351 (PDB ID:7LYQ) [8], Delta Variant B.1.617.2 (PDB ID:7V7V) [34], and Gamma Variant P.1 (PDB ID:7M8K) [32]. All tests follow the same pattern of affinity improvements when using GO linked to heparin compared to heparin alone. This demonstrates that there is a possibility that the compound will be effective against variants, which occurs with most drugs and vaccines available.

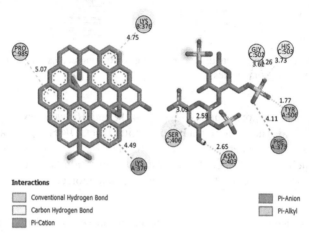

Fig. 5. 2D map for version 01 (torsion4) of the ligand interacting with Spike Protein Omicron Variant (7XNQ)

The interactions of the binding molecules with the amino acids of the protein in version 1 of Table 2 are displayed in Fig. 5, which was created in Biovia Discovery Studio Software [5]. The figure details the protein's amino acids, distances, and types of bonds in the legend. The most common interaction types, namely conventional hydrogen bonding, carbon hydrogen bonding, Pi-cation, Pi-anion, and Pi-alkyl. Asparagine (ASN), lysine (LYS), glycine (GLY), phenylalanine (PHE), serine (SER), proline (PRO), histidine (HIS), tyrosine (TYR) were amino acids.

4   Conclusion

Heparin, which is widely used on a large scale in hospitals for treating various comorbidities and diseases, including COVID-19, exhibited good results in our computer simulations of *ab initio* and molecular docking tests. GO is a graphene-based nanomaterial with functional groups on its surface that facilitates drug targeting and renders the molecule hydrophilic, which is particularly useful in biological media. In this study, GO was functionalized with heparin. After the emergence of the Omicron B.1.1.529 variant of SARS-CoV-2, and other more aggressive variants, they represent a critical challenge to the effectiveness of vaccines and antibody therapies. In our research group, as previously reported, tests were performed with other variants, and our compound maintained a very similar average effectiveness.

First, the high binding power of heparin was confirmed by testing its electronic properties with the GO molecule. Thus, the combination of heparin and GO exhibited a high binding tendency with the spike protein of SARS-CoV-2 virus. This phenomenon indicates a possible gain in viral blocking property, which requires tests for confirmation and analysis of cytotoxicity. Further efficacy tests are ongoing and results are awaited. Finally, molecular dynamics simulations are in progress to confirm the results obtained through molecular docking simulations to help with future practical experiences. Thus, the study is as complete as possible.

Acknowledgments. This work was carried out with the support of the Coordenação de Aperfeiçoamento de Pessoal de Nível Superior-Brazil-CAPES, Financing Code 001, TELEMEDICINA 1690389P, INCT Nanomateriais de Carbono (CNPq). Acknowledgment for computational support from CENAPAD-SP (National Center for High-Performance Processing in São Paulo).

Author's Contributions. AFS, MOM, MZT developed the conception and design of the study. AFS performed computer simulations. AFS reviewed images and wrote the article under the supervision of MOM and SBF. MOM, SBF, MZT reviewed the results. All authors reviewed and commented on the manuscript. All authors approved the final manuscript.

Declaration of Competing Interest. The authors declare that they have no known competing financial interests or personal relationships that could have appeared to influence the work reported in this paper.

References

1. Bursulaya, B.D., et al.: Comparative study of several algorithms for flexible ligand docking. J. Comput. Aided. Mol. Des. **17**(11), 755–763 (2003). https://doi.org/10.1023/B:JCAM.000 0017496.76572.6f
2. Chemcraft: Chemcraft - graphical software for visualization of quantum chemistry computationsy. https://www.chemcraftprog.com. Accessed 05 Dec 2022
3. Clausen, T.M., et al.: SARS-CoV-2 infection depends on cellular Heparan Sulfate and ACE2. Cell **183**(4), 1043-1057.e15 (2020). https://doi.org/10.1016/j.cell.2020.09.033
4. Dacrory, S.: Antimicrobial activity, DFT calculations, and molecular docking of Dialdehyde cellulose/graphene oxide film against Covid-19. J. Polym. Environ. **29**(7), 2248–2260 (2021). https://doi.org/10.1007/s10924-020-02039-5
5. van Dijk, A.D.J., Bonvin, A.M.J.J.: Solvated docking: Introducing water into the modelling of biomolecular complexes. Bioinformatics **22**(19), 2340–2347 (2006). https://doi.org/10.1093/bioinformatics/btl395
6. Fukuda, M., et al.: Lethal Interactions of SARS-CoV-2 with Graphene Oxide: Implications for COVID-19 treatment. ACS Appl. Nano Mater. **4**(11), 11881–11887 (2021). https://doi.org/10.1021/acsanm.1c02446
7. García, A., et al.: SIESTA : recent developments and applications. J. Chem. Phys. **152**(20), 204108 (2020). https://doi.org/10.1063/5.0005077

8. Gobeil, S.M.C., et al.: Effect of natural mutations of SARS-CoV-2 on spike structure, conformation, and antigenicity. Science (1979) **373**, 6555 (2021). https://doi.org/10.1126/science.abi6226

9. Gupta, I., et al.: Antiviral properties of select carbon nanostructures and their functionalized analogs. Mater. Today. Commun. **29**, 102743 (2021). https://doi.org/10.1016/j.mtcomm.2021.102743

10. Gupta, Y., et al.: Heparin: a simplistic repurposing to prevent SARS-CoV-2 transmission in light of its in-vitro nanomolar efficacy. Int. J. Biol. Macromol. **183**, 203–212 (2021). https://doi.org/10.1016/j.ijbiomac.2021.04.148

11. HOPKINS, J.: Coronavirus Resource Center. https://coronavirus.jhu.edu/map.html

12. Hoseini-Ghahfarokhi, M., et al.: Applications of graphene and graphene oxide in smart drug/gene delivery: is the world still flat? Int. J. Nanomed. **15**, 9469–9496 (2020). https://doi.org/10.2147/IJN.S265876

13. ICMAB: Instituto de Ciência de Materiais de Barcelona. https://departments.icmab.es/leem/siesta/

14. Martins, M.O., et al.: Docking fundamentals for simulation in nanoscience. Disciplinarum Scientia - Ciências Naturais e Tecnológicas **22**(3), 67–76 (2021). https://doi.org/10.37779/nt.v22i3.4106

15. Mycroft-West, C.J., et al.: Heparin inhibits cellular invasion by SARS-CoV-2: structural dependence of the interaction of the spike S1 receptor-binding domain with heparin. Thromb. Haemost. **120**(12), 1700–1715 (2020). https://doi.org/10.1055/s-0040-1721319

16. Oliveira, A.M.L., et al.: Graphene oxide thin films with drug delivery function. Nanomaterials **12**(7), 1149 (2022). https://doi.org/10.3390/nano12071149

17. PDB Protein Data Bank: SARS-CoV-2 Omicron BA.4 variant spike. https://www.rcsb.org/structure/7XNQ. https://doi.org/10.1038/s41586-022-04980-y

18. Pedroza, L.S.: Desenvolvimento de novas aproximações para simulações ab initio. USP (2010)

19. Pubchem: Pubchem. https://pubchem.ncbi.nlm.nih.gov/compound/5288499. Accessed 02 Oct 2022

20. Rhazouani, A., et al.: Can the application of graphene oxide contribute to the fight against COVID-19? Antiviral activity, diagnosis and prevention. Curr. Res. Pharmacol. Drug Discov. **2**, 100062 (2021). https://doi.org/10.1016/j.crphar.2021.100062

21. dos Santos, A.F., et al.: In-Silico study of antivirals and non-antivirals for the treatment of SARS-COV-2. Disciplinarum Scientia - Ciências Naturais e Tecno lógicas **23**(2), 57–83 (2022). https://doi.org/10.37779/nt.v23i2.4200

22. Schultz, J.V., et al.: Graphene oxide and flavonoids as potential inhibitors of the spike protein of SARS-CoV-2 variants and interaction between ligands: a parallel study of molecular docking and DFT. Struct. Chem. (2023). https://doi.org/10.1007/s11224-023-02135-x

23. Seabra, A.B., et al.: Nanotoxicity of graphene and graphene oxide. Chem. Res. Toxicol. **27**(2), 159–168 (2014). https://doi.org/10.1021/tx400385x

24. Seifi, T., Reza Kamali, A.: Antiviral performance of graphene-based materials with emphasis on COVID-19: a review. Med. Drug Discov. **11**, 100099 (2021). https://doi.org/10.1016/j.medidd.2021.100099

25. Shafiee, A., et al.: Graphene and graphene oxide with anticancer applications: challenges and future perspectives. MedComm. (Beijing) **3**(1), e118 (2022). https://doi.org/10.1002/mco2.118

26. Tandon, R., et al.: Effective inhibition of SARS-CoV-2 entry by heparin and enoxaparin derivatives. J. Virol. **95**, 3 (2021). https://doi.org/10.1128/JVI.01987-20

27. Trott, O., Olson, A.J.: AutoDock Vina: Improving the speed and accuracy of docking with a new scoring function, efficient optimization, and multithreading. J. Comput. Chem. NA-NA (2009). https://doi.org/10.1002/jcc.21334

28. Unal, M.A., et al.: Graphene oxide Nanosheets interact and interfere with SARS- CoV-2 surface proteins and cell receptors to inhibit infectivity. Small **17**(25), 2101483 (2021). https://doi.org/10.1002/smll.202101483

29. Valdés-Tresanco, M.S., et al.: AMDock: a versatile graphical tool for assisting molecular docking with Autodock Vina and Autodock4. Biol. Direct **15**(1), 12 (2020). https://doi.org/10.1186/s13062-020-00267-2

30. Wang, D., et al.: An overview of the safety, clinical application and antiviral research of the COVID-19 therapeutics. J. Infect. Public Health **13**(10), 1405–1414 (2020). https://doi.org/10.1016/j.jiph.2020.07.004

31. Wang, J., et al.: The inhibition of SARS-CoV-2 3CL Mpro by graphene and its derivatives from molecular dynamics simulations. ACS Appl. Mater. Interfaces **14**(1), 191–200 (2022). https://doi.org/10.1021/acsami.1c18104

32. Wang, P., et al.: Increased resistance of SARS-CoV-2 variant P.1 to antibody neutralization. Cell Host Microbe **29**(5), 747–751.e4 (2021). https://doi.org/10.1016/j.chom.2021.04.007

33. Wang, X., et al.: Structural insights into the cofactor role of Heparin/Heparan Sulfate in binding between the SARS-CoV-2 spike protein and host Angioten sin-converting enzyme II. J. Chem. Inf. Model. **62**(3), 656–667 (2022). https://doi.org/10.1021/acs.jcim.1c01484

34. Wang, Y., et al.: Structural basis for SARS-CoV-2 delta variant recognition of ACE2 receptor and broadly neutralizing antibodies. Nat. Commun. **13**(1), 871 (2022). https://doi.org/10.1038/s41467-022-28528-w

Make No Mistake! Why Do Tools Make Incorrect Long Non-coding RNA Classification?

Alisson G. Chiquitto[1,3], Lucas Otávio L. Silva[1], Liliane Santana Oliveira[1], Douglas S. Domingues[1,2], and Alexandre R. Paschoal[1(✉)]

[1] Department of Computer Science, Bioinformatics and Pattern Recognition Group, Federal University of Technology - Paraná - UTFPR, Cornélio Procópio, PR, Brazil
paschoal@utfpr.edu.br
[2] Department of Genetics, "Luiz de Queiroz" College of Agriculture, Group of Genomics and Transcriptomes in Plants, University of São Paulo, Piracicaba, SP, Brazil
[3] Federal Institute of Education, Science and Technology of Mato Grosso do Sul - IFMS, Campus Navirai, MS, Brazil

Abstract. Long non-coding RNAs (lncRNAs) play important roles in various biological processes, and their accurate identification is essential for understanding their functions and potential therapeutic applications. In a previous study, we assessed the impact of short and long reads sequencing technologies on long non-coding RNA computational identification in human and plant data. We provided evidence of where and how to make potential better approaches for the lncRNA classification. In this follow-up study, we investigate the misclassified sequences by five machine learning tools for lncRNA classification in humans to understand the reasons behind the failures of the tools. Our analysis suggests that the primary cause for the failures of these tools is the overlap of two coding regions by lncRNAs, similar to a chimeric sequence. Furthermore, we emphasize the need to view genes as transcriptional units, as the transcript will define the gene function. These insights underscore the need for further refinement and improvement of these tools to enhance their accuracy and reliability in lncRNA prediction and classification, ultimately contributing to a better understanding of the role of lncRNAs in various biological processes and potential therapeutic applications.

Keywords: Non-coding RNAs · high-throughput sequencing technologies · coding · methods · benchmarking

1 Introduction

Long-read sequencing provides numerous advantages over short-read sequencing [1,12]. While short-read sequencing produces reads that are typically less than 300 bases, long-read sequencing technologies generate reads that can be more

M. S. Reis and R. C. de Melo-Minardi (Eds.): BSB 2023, LNBI 13954, pp. 36–45, 2023.
https://doi.org/10.1007/978-3-031-42715-2_4

than 10 kb in length [12]. Long reads can thus improve *de novo* assembly, mapping certainty, transcript isoform identification, and detection of structural variants. Additionally, long-read sequencing of native molecules, such as DNA and RNA, eliminates amplification bias and preserves base modifications. As accuracy, throughput, and cost reduction continue to improve, long-read sequencing is becoming an increasingly attractive option for a wide range of genomics applications, including those for model and non-model organisms [1,15].

Long non-coding RNA (lncRNA) identification is a significantly benefited class of transcripts that would have improved identification using long-read sequencing technologies [10]. The GENCODE project [8], as a foundational resource for human genome annotation, has recently recognized the benefits of using long-read transcriptome sequencing [8]. In the human transcriptome, the benefits of annotating lncRNAs using long-read transcript data have already been proven to be an important advance (e.g., some reads span the entire transcript and do not require assembly) [14]. The latest versions of the lncRNA annotation in the GENCODE database are massively based on long-read sequencing, making this database a reference resource for evaluating the performance of lncRNA computational predictors.

In our previous work [2], we examined the impact of sequencing technologies on lncRNA prediction using human genome annotation. We obtained lncRNA transcripts datasets from the GENCODE project in releases 21 (V21) and 38 (V38), containing 26,414 and 48,751 lncRNA sequences respectively. Release V21 relied mostly on short-sequencing data, while V38 predominantly used long-read data. These lncRNA datasets served as inputs for 10 prediction tools, and their performance was evaluated based on sensitivity. We then selected the five tools with the best average sensitivity across V21 and V38 - RNAmining, PLEK, CNCI, LncADeep, and lncRNAnet - for a more detailed lncRNA analysis. Following our subsequent investigation, we found nine misclassified sequences for V21 and five for V38 across the five tools (see details in [2])

In this report, we focus specifically on misclassification (misclassified lncR-NAs), which occurs when an input that is a true lncRNA is incorrectly classified as coding. We first updated our sensitivity analysis to GENCODE version 41 (V41) and compared it with our previous report [2]. We then investigated in detail the transcripts that were commonly incorrectly classified by all the tools we analyzed. We discuss potential reasons for misclassification that could inform the improvement of machine learning models. Our study has yielded two significant findings that are of relevance contributions: firstly, identifying lncRNAs that overlap with multiple coding genes, which behave like chimeras, poses a challenge for accurate identification; and secondly, the traditional definition of gene as either coding or non-coding is no longer applicable as the transcript itself plays a critical role in determining its function. Our aim is to gain a better understanding of why these tools make mistakes in lncRNA classification. Understanding the reasons behind lncRNA misclassification is crucial for improving the accuracy of lncRNA classification tools and advancing our understanding of lncRNA biology.

2   Methods

We selected the five top-performing tools - RNAmining, PLEK, CNCI, LncADeep, and lncRNAnet - with the best average values based on our previous report [2]. We focused specifically on misclassification, where the input data is a true lncRNA but is incorrectly classified as coding. All five tools utilize supervised machine learning methods to differentiate between lncRNAs and protein-coding genes (mRNA) or between coding and non-coding sequences. To perform our analysis, we used 54,291 lncRNA transcript sequences from GEN-CODE version 41 (V41) and compared the results with previous analyses in versions 21 (V21) and 38 (V38). We updated the sensitivity analysis of each tool for V41 to ensure the most accurate results. We used the Ensembl Genome Broswer visualization from GENCODE version 43 which keep the same results as V41.

The sensitivity is a metric that measures the accuracy of positive predictions and is calculated as the ratio of the total number of correct positive predictions to the number of actual positives. In our study, we used lncRNA transcripts as the positive set and evaluated the sensitivity of each tool by comparing the True Positives (TP) and False Negatives (FN) counts. The sensitivity metric was calculated using the Eq. 1, where TP represents the correct classifications made by the tools and FN represents the number of lncRNA transcripts misclassified as coding transcripts. In the other words, the FN meaning sequences incorrectly identified as coding sequences when they should have been classified as lncRNAs according to GENCODE.

$$SN = TP/(TP + FN) \tag{1}$$

Finally, we used Upset plots and Venn diagrams (Fig. 2a) to visually represent the misclassified results by the tools.

3   Results and Discussion

In our previous report [2], we evaluated ten lncRNA classification tools in the GENCODE V21 dataset (short read data) against V38 datasets (long read data). We identified the five tools with the highest average sensitivity in comparison to sequencing technologies: RNAmining (99.22%), LncADeep (97.13%), CNCI (97.47%), lncRNAnet (96.91%), and PLEK (95.99%) (Fig. 1).

After comparing the results obtained from the V21 and V38 datasets, we observed that, except for PLEK, all other tools demonstrated improved performance. This led us to conclude that the V38 dataset, which is based on long-read sequencing technology, was better annotated. One possible explanation for this observation is that PLEK relies solely on the k-mer scheme (k = 1..5) to characterize lncRNA and mRNA transcriptional sequences, whereas the other tools employ at least two strategies for feature extraction.

In this report, we ran the same five tools with the V41 dataset as input and updated their performance based on sensitivity values (Fig. 1). Our analysis revealed the following order of tools based on their sensitivity values: RNAmining (99.47%), LncADeep (97.66%), CNCI (97.49%), lncRNAnet (97.23%), and PLEK (92.88%). These findings are consistent with our earlier evaluation of the five tools using the V38 dataset [2].

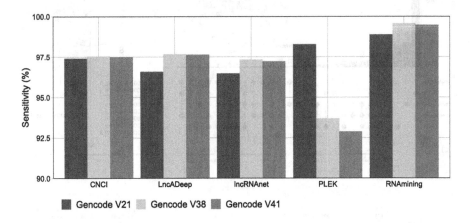

Fig. 1. Comparison of sensitivity values for five lncRNA classification tools (CNCI, LncADeep, lncRNAnet, PLEK, and RNAmining) across three GENCODE datasets: V21, V38, and V41. The sensitivity values were calculated based on the ability of each tool to accurately classify lncRNAs in each dataset. The y-axis represents the sensitivity values, while the x-axis shows each tool and bars each dataset used as input. The results show that RNAmining consistently outperformed the other tools across all three datasets, while PLEK showed the lowest sensitivity, especially in the V41 dataset. The findings are consistent with our earlier evaluation of these tools on the V38 dataset [2].

We also determined the number of sequences misclassified by more than one tool. A total of 6,249 sequences were misclassified by at least one tool, and that 5,199 (83.2%) sequences were misclassified exclusively by one tool (see Upset Plot in Fig. 2a). The intersection of all sets (in red in Fig. 2a) shows that only five sequences were misclassified by all five tools used in this study (V41). PLEK, in particular, had the highest number of misclassifications, with 3,557 (56,92% of total) sequences classified incorrectly, of which 3,075 (59.15% of exclusives) were exclusively misclassified by PLEK.

In the previous study, we also compared the misclassified sequences and found that nine transcripts in V21 and five in V38 were wrongly classified by all five tools used. Using this information, we compared the misclassified transcripts for each of the GENCODE releases (V21, V38, and V41), and we found that the union of all versions resulted in 14 distinct misclassified transcripts (Fig. 2b). Among the eight misclassified sequences exclusively from V21, four transcripts were annotated as protein-coding in V41, and the other four no longer have transcript annotation information, indicating an improvement in transcript annotation. The one misclassified exclusive sequence from V38 (ENST00000675965) is

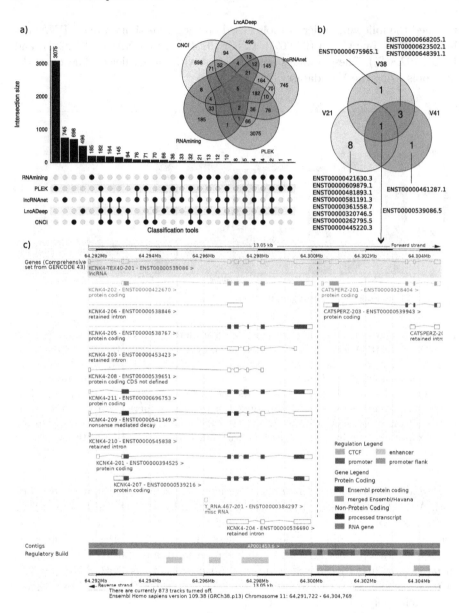

Fig. 2. (a) Misclassified sequences by the five tools in V41. The rows correspond to a set of misclassified sequences. Each column corresponds to a possible intersection: the filled-in circles show which set is part of an intersection, and vertical bar charts show the size of the intersection. Five sequences were identified as non-lncRNA by five lncRNA tools in V41. (b) The intersection of misclassified sequences by five lncRNA tools. Nine sequences were misclassified in V21, five in V38, and five in V41. Transcript IDs are displayed around the diagram. We show the latest version of the transcripts. (c) Ensembl Genome Browser showing the genetic region and isoforms of the ENST00000539086 transcript. The vertical dashed line is between the two coding genes overlapped by the lncRNA (adapted from [4]).

also not present in V41. The unique exclusive misclassified sequence from V41 (ENST00000461287) is a novel lncRNA transcript that was previously defined as a coding sequence.

We found a transcript (ENST00000539086) that was misclassified in all three versions of GENCODE (Fig. 2c). Upon checking the Ensembl Browser, we discovered that there are 12 isoforms within this transcript, with only one as an lncRNA. However, the critical aspect is that this lncRNA isoform overlaps with two coding genes, making it look like a chimera. This evidence indicates that regions with coding and non-coding features make their distinction difficult [9], and we believe this is the reason for the misclassification of this transcript.

Another important factor to consider is the potential for region overlap. Typically, lncRNA classification tools are trained using two separate sets of transcripts: one consisting of mRNAs, and the other of lncRNAs. This means that when a lncRNA overlaps like a chimera with an mRNA, there is a chance that a transcript (or a significant portion of it) may be present in both training sets of the lncRNA classification tool. Consequently, this intersection of data can compromise the accuracy of the classifier.

The transcripts ENST00000648391 and ENST00000461287 exhibit the same chimera situation as discussed previously for transcript ENST00000539086. It is hypothesized that this similarity could lead to failures in classifying these transcripts by the tools. Figures 3 and 4 show these overlapping regions.

There are only three misclassified lncRNA transcripts in common between V38 and V41 (ENST00000668205, ENST00000623502, and ENST00000648391). Two (ENST00000668205 and ENST00000623502) of them are lincRNA transcripts that overlap two enhancer regions and multiple sequence regions (e.g.,

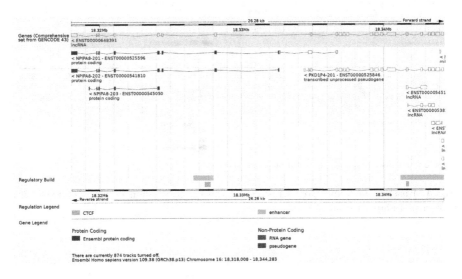

Fig. 3. Ensembl Genome Browser showing the genetic region of ENST00000648391 transcript (adapted from [6])

enhancers and promoter regions). Figures 5 and 6 show the region of these transcripts. The third is a novel lncRNA transcript (ENST00000648391) that does not have an identical model in RefSeq according to the Ensembl Human (GRCh38.p13) website. According to the Ensembl annotation, this transcript has a length of 4082 bps and 22 associated exons and we already described before (Fig. 3).

Fig. 4. Ensembl Genome Browser showing the genetic region of ENST00000461287 transcript (adapted from [3])

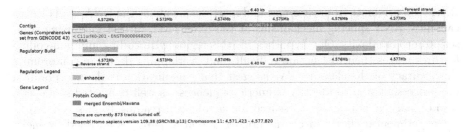

Fig. 5. Ensembl Genome Browser showing the genetic region of ENST00000668205 transcript (adapted from [7])

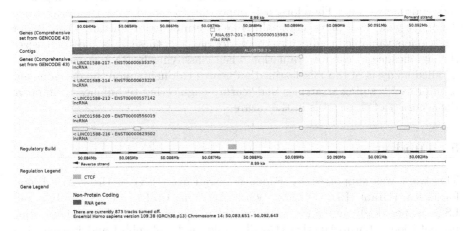

Fig. 6. Ensembl Genome Browser showing the genetic region of ENST00000623502 transcript (adapted from [5])

4 Conclusions

In this study, we thoroughly evaluated the performance of five machine learning tools (RNAmining, LncADeep, CNCI, lncRNAnet, and PLEK) for classifying long non-coding RNA (lncRNA) sequences in three versions of the GENCODE dataset (V21, V38, and V41). Through our analysis, we aimed to gain a deeper understanding of why these tools can make mistakes in lncRNA classification. To gain insights into the specific types of misclassifications, we analyzed the individual sequences that were misclassified by each tool. Interestingly, we found that most misclassifications occurred when a sequence contained features of both coding and non-coding RNA, making it difficult to distinguish between the two. In particular, we identified a transcript (ENST00000539086) misclassified in all three GENCODE versions, with 12 isoforms within the transcript and only one as an lncRNA, overlapping with two coding genes, making it look like a chimera. Chimeric RNAs, as described by [13], are formed through gene fusion events, contributing to the intricacy of the transcriptome. These chimeric

RNAs can originate from chromosomal rearrangements at the DNA level or non-canonical RNA splicing mechanisms at the RNA level, further enhancing transcriptome complexity. Our hypothesis is this region with coding and non-coding features makes their distinction difficult, and overlapping data can compromise the accuracy of the classifier. We suggest that classification tools incorporate a pre-processing step to identify chimeric sequences, as well truncated sequences. Additionally, it is possible to search for small open reading frames (ORFs) to aid experimental efforts in characterizing potential peptides encoded by lncR-NAs (as done by [11]). Another key conclusion is that the traditional definition of a gene as either coding or noncoding no longer applies, since a transcript can be both simultaneously. It is the transcript that determines whether a gene is coding or non-coding. Therefore, it is important to consider the transcript when defining a gene, as it plays a crucial role in determining its function. Overall, our study offers valuable insights into the performance of these tools and emphasizes the significance of careful evaluation when classifying lncRNA sequences. Our detailed analysis of the misclassified sequences can help guide the development of improved training strategies and feature engineering techniques to improve the accuracy of these tools in the future.

5   Funding

This work was supported by Fundação Araucária (FA) in the NAPI Bioinformática (Grant PDI 66/2021 with ARP as Co-PI, LOLS as IC, and LSO as postdoc); and STIC AmSud (https://www.sticmathamsud.org/stic/presentacion/) Latin America (Brazil, Chile, and Colombia) and France from TELearning Project 2021-22 (21-STIC-13). ARP thanks to NVIDIA from the GPU Grant Program 2019 - Accelerated Data Science Call for the GPU Seed Units. AGC is a Ph.D. Student and he was supported from the Ph.D. Programme in Bioinformatics (PPGAB - UTFPR and UFPR). LOLS received scientific initiation fellowship from PROPPG/UTFPR/CNPq. DSD research on plant transcriptome analysis are funded by CNPq (312823/2019-3) and FAPESP (2016/10896-0, 2018/08042-8 and 2019/15477-3).

References

1. Burgess, D.J.: Genomics: next regeneration sequencing for reference genomes. Nat. Rev. Genet. **19**(3), 125 (2018)
2. Chiquitto, A.G., Silva, L.O.L., Oliveira, L.S., Domingues, D.S., Paschoal, A.R.: Impact of sequencing technologies on long non-coding RNA computational identification. In: 2022 IEEE International Conference on Bioinformatics and Biomedicine (BIBM) (2022)
3. Ensembl: Ensembl genome browser enst00000461287 (2023). www.feb2023. archive.ensembl.org/Homo_sapiens/Share/ba38b47c75f9e62e9cd82253bdcc235b? redirect=no. Accessed 31 Mar 2023

4. Ensembl: Ensembl genome browser enst00000539086 (2023). www.feb2023. archive.ensembl.org/Homo_sapiens/Share/1a6c08c69bf3fcfb9494fcbb2d1676cb? redirect=no. Accessed 31 Mar 2023

5. Ensembl: Ensembl genome browser enst00000623502 (2023). www.feb2023. archive.ensembl.org/Homo_sapiens/Share/7e058f22d6c8e5c849c29b7be72fd5a0? redirect=no. Accessed 31 Mar 2023

6. Ensembl: Ensembl genome browser enst00000648391 (2023). www.feb2023. archive.ensembl.org/Homo_sapiens/Share/a66509fa49c2933d0c22da068b44c2c2? redirect=no. Accessed 31 Mar 2023

7. Ensembl: Ensembl genome browser enst00000668205 (2023). www.feb2023. archive.ensembl.org/Homo_sapiens/Share/3d5e32afaa48f26431ba59ae949b68d9? redirect=no. Accessed 31 Mar 2023

8. Frankish, A., et al.: GENCODE 2021. Nucleic Acids Res. **49**(D1), D916–D923 (2020). https://doi.org/10.1093/nar/gkaa1087

9. Klapproth, C., Sen, R., Stadler, P.F., Findeiß, S., Fallmann, J.: Common features in lncRNA annotation and classification: a survey. Non-Coding RNA **7**(4), 77 (2021). https://doi.org/10.3390/ncrna7040077

10. Lagarde, J., et al.: High-throughput annotation of full-length long noncoding RNAs with capture long-read sequencing. Nat. Genet. **49**(12), 1731–1740 (2017). https:// doi.org/10.1038/ng.3988. www.nature.com/articles/ng.3988

11. Nabi, A., Dilekoglu, B., Adebali, O., Tastan, O.: Discovering misannotated lncR-NAs using deep learning training dynamics. Bioinformatics **39**(1) (2023). https:// doi.org/10.1093/bioinformatics/btac821

12. Pollard, M.O., Gurdasani, D., Mentzer, A.J., Porter, T., Sandhu, M.S.: Long reads: their purpose and place. Hum. Mol. Genet. **27**(R2), R234–R241 (2018). https:// doi.org/10.1093/hmg/ddy177

13. Wang, Y., et al.: Identification of the cross-strand chimeric RNAs generated by fusions of bi-directional transcripts. Nat. Commun. **12**(1), 4645 (2021). https:// doi.org/10.1038/s41467-021-24910-2

14. Xie, S.Q., et al.: ISOdb: a comprehensive database of full-length isoforms generated by Iso-Seq. Int. J. Genomics **2018**, 1–6 (2018) https://doi.org/10.1155/2018/ 9207637. www.hindawi.com/journals/ijg/2018/9207637/

15. Yuan, Y., Bayer, P.E., Batley, J., Edwards, D.: Improvements in genomic technologies: application to crop genomics. Trends Biotechnol. **35**(6), 547–558 (2017)

Spectrum-Based Statistical Methods for Directed Graphs with Applications in Biological Data

Victor Chavauty Villela, Eduardo Silva Lira, and André Fujita[✉]

Instituto de Matemática e Estatística, Universidade de São Paulo (USP),
São Paulo, Brazil
andrefujita@usp.br

Abstract. Graphs often model complex phenomena in diverse fields, such as social networks, connectivity among brain regions, or protein-protein interactions. However, standard computational methods are insufficient for empirical network analysis due to randomness. Thus, a natural solution would be the use of statistical approaches. A recent paper by Takahashi et al. suggested that the graph spectrum is a good fingerprint of the graph's structure. They developed several statistical methods based on this feature. These methods, however, rely on the distribution of the eigenvalues of the graph being real-valued, which is false when graphs are directed. In this paper, we extend their results to directed graphs by analyzing the distribution of complex eigenvalues instead. We show the strength of our methods by performing simulations on artificially generated groups of graphs and finally show a proof of concept using concrete biological data obtained by Project Tycho.

Keywords: Network Correlation · Graph Statistics · ECoG

1 Introduction

We often use graphs to model interactions between objects. Some examples include the functional connectivity of brain regions [4], social interactions [18], molecular interactions [2], and gene regulations [1].

Once we model these natural phenomena using graphs, it becomes of significant interest to discriminate graphs of two or more populations or make inferences [10]. For instance, suppose three patient groups were assigned different treatments for a neurochemical condition. By examining each patient's resting state magnetic resonance imaging (MRI) scans, can we discern whether there is a notable distinction among the administered drugs?

Traditional computation methods rely on the search for an isomorphism between graphs or sub-graphs, which are prone to failure when randomness is

This study was financed in part by the Coordenação de Aperfeiçoamento de Pessoal de Nível Superior - Brasil (CAPES) - Finance Code 001.

M. S. Reis and R. C. de Melo-Minardi (Eds.): BSB 2023, LNBI 13954, pp. 46–57, 2023.
https://doi.org/10.1007/978-3-031-42715-2_5

applied to the graphs [9]. Because of this nature, these methods are unfit for usage in biological data, where intrinsic randomness is expected [10].

An alternative technique is to compare graph features, such as the number of nodes and edges, and, particularly, centrality measures, such as closeness and betweenness [8]. These centrality measures are estimated and then used as input in standard statistical methods. Although this is a step up from the previous techniques, centrality measures can under-represent variability between graphs. Take, for example, two graphs obtained from the Watts-Strogatz model. Even if distinct rewiring probabilities are used, they still present the same centrality measure since the number of edges does not change [10].

In 2017, Takahashi *et al.* [12] proposed that the graph spectrum is a good feature for describing the graph structure. They used the Kullback-Leibler and Jensen-Shannon divergences between spectral distributions to measure the distance between graphs. Using this concept, they constructed tools for 1. model selection; 2. a parameter estimator for random graph models; 3. a statistical test to compare two sets of graphs. More recently, these ideas have been used to create a concept of correlation [11]/causality [15] between graphs and spectrum-based clustering algorithms for complex networks [14].

One limitation of this work is that it is limited to undirected graphs whose eigenvalues are all real-valued. However, many empirical graphs are directed. A solution would be to symmetrize the graph. The problem is that we usually lose the directionality information, or it vastly influences the spectrum distribution.

In this paper, we extend the results of Takahashi et al. for directed graphs. Our ANOVA-like approach can distinguish between groups of directed graphs obtained from distinct populations. Also, we apply it to actual biological data for illustration.

2   Materials

2.1   Graphs

A graph G consists of a pair (N, E), where N is a set of nodes, and E is a set of edges connecting a pair of nodes of G.

We call a graph weighted if every edge between two nodes i and j is associated with a complex value $e_{i,j} \in \mathbb{C}$. In contrast, in non-weighted graphs, an edge between two nodes i and j will assume 1 if i and j are connected or 0 otherwise.

A graph is said to be undirected if, for every pair of nodes i and j, the edges $e_{i,j}$ and $e_{j,i}$ connecting i to j, and j to i respectively, are equal. That is: $e_{i,j} = e_{j,i}$. Otherwise, it is undirected.

Given a graph G with n nodes, we define its adjacency matrix $\mathbf{A}_G$ as the matrix $\mathbf{A}_G = (e_{i,j})_{i,j=1,...,n}$, where $e_{i,j}$ is the value associated with the edge connecting node i and node j. Note that the adjacency matrix of an undirected graph is symmetric.

The spectrum of a graph G is the set of eigenvalues of its adjacency matrix $\mathbf{A}_G$. If G is directed, its adjacency matrix is non-symmetrical. Therefore its eigenvalues are complex-valued.

2.2   Spectral Distribution

A random graph g is a family of graphs whose members are generated by some probability law. For example, we construct an Erdös-Rényi random graph by connecting two nodes with probability p.

We define the complex Dirac delta as the measure $\delta_\mathbb{C}$ satisfying for every compactly supported continuous function f:

$$\int_\mathbb{C} f(x)\delta_\mathbb{C}\{dx\} = f(0).$$

Alternatively, we construct the complex Dirac delta function as the product of the 1-dimensional Dirac delta in two variables (the real and the imaginary variables). That is:

$$\delta_\mathbb{C}(a + bi) = \delta(a)\delta(b).$$

Let g be a directed random graph generated by some probability law. Then its complex eigenvalues Δ form random vectors. Let brackets $\langle\rangle$ indicate expectations concerning the probability law. Then we define the spectral distribution of the directed random graph g as

$$\rho_g(\lambda) = \lim_{n\to\infty} \langle \frac{1}{n} \sum_{j=1}^{n} \delta_\mathbb{C}(\lambda - \frac{\lambda_j}{\sqrt{n}}) \rangle.$$

This distribution is highly correlated with distinct features of the graph. We can use it as a fingerprint of the random graph g [10].

2.3   Calculating the Graph Spectrum

Estimating the spectral distribution of a directed random graph is performed under a similar procedure as for the undirected case [10].

Since the spectral density ρ_g is unknown, we need an estimator $\hat{\rho}_g$. We initially compute the eigenvalues $\lambda_1, \ldots, \lambda_n$ of the graph's adjacency matrix g and apply a multivariate kernel regression [6]. We divide the resulting 2-dimensional surface by the volume under the curve to ensure the final volume is one (probability function).

2.4   Statistical Distance

The spectrum distribution is the distribution of complex eigenvalues of a graph model. We are interested in using it as a fingerprint of the model so that by comparing the spectrum of two different random graph models, we can establish a certain distance between them. Similarly, we can compare the spectrum of a graph to the spectrum distribution of a random graph model and obtain a measure of how far apart the graph is from being generated from that specific model.

To compare these distributions, we will be using the Kullbeck-Leibler divergence [13]. The Kullbeck-Leibler divergence is a statistical distance measuring how a probability distribution differs from a second distribution. For two probability densities p and q, the Kullbeck-Leibler divergence is defined as

$$D(p, q) = \int_{\mathbb{C}} p(x) \log \left(\frac{p(x)}{q(x)} \right) dx$$

2.5   Random Graph Models

Graphs can often model very complex phenomena, and it is often impossible to establish how a graph was formed when dealing with biological data. Besides, it is difficult to establish whether two graphs are similar simply by analyzing their structures. Thus, one idea is to imagine these graphs resulting from a probabilistic model with a set of parameters.

Directed Models. Unfortunately, models for directed graphs are not as prevalent as the ones for undirected graphs. Thus, we propose the following general extension of any directed model.

Given a random model r with a parameter p, we extend this model as follows. Let p_1 and p_2 be two parameters for model r. Then

1. Generate a graph G_1 with parameter p_1 and construct its adjacency matrix.
2. Generate a graph G_1 with parameter p_2 and construct its adjacency matrix.
3. Generate a matrix M whose upper triangular is the same as of G_1 and whose lower triangular is the same as of G_2.
4. Generate a graph G with adjacency matrix M.

Note that the parameters p_1 and p_2 control the network's inner and external connections, respectively, which are represented on the upper and lower triangle of the graph's adjacency matrix. In the scenario in which $p_1 = p_2$, the resulting graph is still directed due to the random element of the graph generation process.

We will use this procedure to run our simulations.

3   Methods

Given k groups of graph samples, we are now interested in verifying whether they originated from the same population.

Naively, we could use a parametric approach by selecting a random graph model, estimating the parameters for each graph, and using traditional ANOVA with the estimated parameters as input. However, we must know the random graph model, which is very unlikely in most realistic scenarios. Other nonparametric methods, like the Kolmogorov-Smirnov test, require independence of the graphs, which is often not true when they result from a biological process. Therefore, we will use an ANOVA-like approach following the ideas described by Fujita *et al.* [9] called ANOGVA.

In other words, we will perform a variation of the ANOVA using the complex distribution of eigenvalues of the graphs.

Let $g_1, \ldots, g_k$ be k distinct graph populations. If these graphs come from the same population, their spectral distributions should be equal. Let ρ_{g_i} be the average graph spectrum for group i, $\rho_G = \frac{1}{k} \sum_{i=1}^{k} \rho_{g_i}$ be the overall graph spectrum average, and D be the Kullbeck-Leibler divergence.

Then, we test the following hypothesis:

$$H_0 : D(\rho_{g_1}, \rho_G) = D(\rho_{g_2}, \rho_G) = \ldots = D(\rho_{g_k}, \rho_G) = 0$$

H_1: At least one of the groups of graphs was generated in a different manner

Under the null hypothesis, we expect the statistic $\Delta = \sum_{i=1}^{k} D(\rho_{g_1}, \rho_G)$ to be small. Under the alternative hypothesis, we expect it to be large.

The distribution of Δ is unknown and highly dependent on the used random graph model. Therefore, to test for significance, we will use a bootstrap approach.

The following algorithm describes how we compute the bootstrap

Input: k groups of graphs, $g_1, \ldots, g_k$, and a number of max-iterations
        Max
Output: A p-value
1 Estimate $\hat{\rho}_{g_1}$ and $\hat{\rho}_G$;
2 Calculate $\hat{\Delta} = \sum_{i=1}^{k} D(\hat{\rho}_{g_1}, \hat{\rho}_G)$;
3 Set $\hat{\Delta}_l = []$;
4 **for** Max *iterations* **do**
5 | Construct k new groups $g'_1, \ldots, g'_k$ by resampling (without
    | replacement) the original graph set;
6 | Estimate the average spectrum distribution $\hat{\rho}_{g'_i}$ for each new graph
    | g'_i;
7 | Calculate the overall graph spectrum average $\hat{\rho}'_G$;
8 | Calculate $\hat{\Delta}' = \sum_{i=1}^{k} D(\hat{\rho}'_{g_1}, \hat{\rho}'_G)$.;
9 | Append $\hat{\Delta}'$ to $\hat{\Delta}_l$;
10 **end**
11 Let $p = \mathbf{Cardinality}(\hat{\Delta}' \in \hat{\Delta}_l : \text{such that } \hat{\Delta}' \geq \hat{\Delta}) \cdot \frac{1}{Max}$;
12 **return** p;

Algorithm 1: Anogva

Implementation. We implemented this method in R, extending the existing StatGraph package [17]. We constructed the multivariate kernel density estimator using the package 'ks.'

4   Simulations

To verify the power of the method described in this paper, we constructed a set of simulations to generate directed random graphs as defined in Sect. 2.5.

To evaluate the performance of ANOGVA, we need to verify the null (H_0) and alternative (H_1) hypotheses. Since we want to ensure that ANOGVA works in a wide range of random graph models, we generated the graphs using the Erdös-Rényi [7], Watts-Strogatz [19], and Barabási-Albert [3] models, as described in Sect. 2.4. We generated the graphs using the igraph package in R [5].

For each of the models, we performed the following simulation:

1. We generated three sets of graphs: G_1, G_2, and G_2, each containing ten graphs, for a total of 30 graphs.
2. All of the graphs were generated with $n = 800$ nodes and using specific parameters.
3. We then applied the ANOGVA algorithm using 500 bootstrap samples.
4. We ran this experiment 500 times, generating a p-value distribution.

Simulation (H_0): All three groups should be generated using the same set of parameters under the null hypothesis.

Table 1 describes the parameters we used for each random graph model.

Table 1. Parameters used in the null hypothesis simulation

Model	Parameters
Erdös-Rényi	$p_1 = 0.1$ and $p_2 = 0.2$
Watts-Strogatz	$p_1 = 0.1$ and $p_2 = 0.3$ $neigh = 10, dim = 1$
Barabási-Albert	$p_1 = 1.0$ and $p_2 = 1.1$

Since we generated all groups using the same models and parameters, we can safely assure that they all come from the same population. In other words, they are under the null hypothesis. Under the null hypothesis, we expect that the distribution of p-values forms a uniform distribution in the $[0, 1]$ range.

Figure 1 shows the distribution of the p-values. As we can see, they form a uniform distribution, thus showing that our proposal controls the rate of false positives.

We remark that the power of the test increases with the number of graphs. Thus, a small number of graphs ($N = 30$) shows that even under a small sample size, the ANOGVA method performs well.

Now we verify the H_1 hypothesis.

Simulation (H_1): To verify the power of the test, we need to generate groups from distinct populations.

Table 2 describes which parameters we used for each random graph model.

Since we generated all the groups using distinct parameters, this simulation satisfies the requirements for H_1, where we generated at least one of the groups (in this case, all of them) differently. Under the alternative hypothesis, we expect all the p-values to be small.

In all models, the resulting p-values were all equal to zero.

We can see that the ANOGVA method satisfies our expectations.

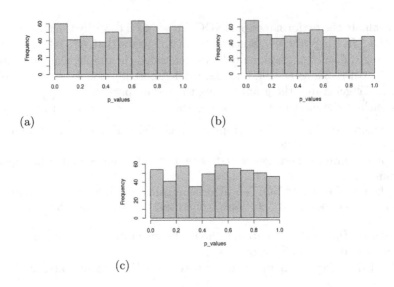

(a)   (b)

(c)

Fig. 1. (a) Distribution of p-values for the ANOGVA simulation under the null hypothesis using the Erdős-Rényi model. (b) Distribution of p-values for the ANOGVA simulation under the null hypothesis using the Watts-Strogatz model (c) Distribution of p-values for the ANOGVA simulation under the null hypothesis using the Barabási-Albert model. Notice that all of them are uniform distributions. Performing a Kolmogorov-Smirnov test comparing these values with the uniform distribution gives us p-values greater than 0.05.

Table 2. Parameters used in the alternative hypothesis simulations

Model	G_1	G_2	G_3
Erdős-Rényi	$p_1 = 0.1, p_2 = 0.3$	$p_1 = 0.2, p_2 = 0.4$	$p_1 = 0.3, p_3 = 0.2$
Watts-Strogatz	$p_1 = 0.1, p_2 = 0.3, neigh = 10, dim = 1$	$p_1 = 0.2, p_2 = 0.7, neigh = 10, dim = 1$	$p_1 = 0.3, p_3 = 0.3, neigh = 10, dim = 1$
Barabási-Albert	$p_1 = 1.1, p_2 = 1.3$	$p_1 = 1.1, p_2 = 1.8$	$p_1 = 1.7, p_2 = 1.8$

5   Applications to Biological Data

To illustrate ANOGVA, we applied it to a biological dataset. We used the data source titled 'Anesthesia Task' [20]. We obtained it from Project Tycho and downloaded it via their website at http://wiki.neurotycho.org The experiment aimed to compare neural activity between most of the lateral cortex measured with electrocorticographic signals (ECoG) in a macaque during five stages: awake with eyes opened, awake with eyes closed, anesthetized, recovering with eyes closed, and recovering with eyes open.

5.1   Data Source

Four experiments were conducted, each on a different monkey. In each experiment, a monkey was seated in a chair with restricted arms and head movement.

In particular, the following steps describe the experiment for the monkey we ana-lyzed. Neural data was acquired through 128 ECoG electrodes measuring ECoG signals from most of the lateral cortex. Neural activity was recorded during all of the following stages. Initially, the monkey was awake and opened its eyes, sitting calmly in his chair for 10 min. Next, the eyes of the monkey were covered with an eye mask to avoid evoking a visual response. The monkey was left sitting in his chair for another 10 min. Recording of neural activity was stopped while anesthesia was intramuscularly injected into the monkey. By the point at which the monkey had stopped responding to manipulation of the monkey's hand or touching the nostril or philtrum with a cotton swab, neural activity recording was resumed for another 20 min. After the anesthetized condition, the monkey recovered from the anesthesia and was left alone for 55 min with its eyes still covered. Next, the eye mask was removed, and the monkey was left to sit calmly on his chair for another 10 min.

5.2   Data Processing and Graph Generation

The initial data generated by the experiment consisted of 128-time series in 5 categories: conscious with open eyes, conscious with closed eyes, anesthetized, recovering with closed eyes, and recovering with open eyes.

Initially, the data was processed through several finite impulse response (FIR) filters to remove any effect caused by electrical interference. We divided the filtered data into several time windows, each lasting four seconds and generated the graph using generalized partial directed coherence (gPDC) [16].

The gPDC is a frequency domain approach to identify the direction of information flow (Granger causality) between multiple time series. We say that a time series X Granger cause another time series Y if knowledge of $X(t-1), \ldots, X(t-k)$ increases the prediction of $Y(t)$.

We carried out gPDC on the 128 frequencies of the filtered data. The result was five sets of 128 groups of graphs (one for each generated frequency). Each group consisted of several graphs, each representing a time window in its cate-gory. Each graph had 128 nodes (each corresponding to a different ECoG elec-trode). The graph was directed and weighted, where each edge between two nodes corresponded to the level of causality between the ECoG electrodes.

5.3   ANOGVA

We performed the following experiment to verify the power of the ANOGVA method. We selected a single-frequency domain. Given that frequency, we chose 100 graphs from each category. This procedure resulted in the following:

1. G_1: 100 graphs generated from when the monkey was awake with its eyes opened.
2. G_2: 100 graphs generated from when the monkey was awake with closed eyes.
3. G_3: 100 graphs generated from when the monkey was under anesthesia.

4. G_4: 100 graphs generated from when the monkey was recovering with closed eyes.
5. G_5: 100 graphs generated from when the monkey was recovering with closed eyes.

First Experiment: We first performed an ANOGVA test using the five groups. We used 1 000 bootstrap samples.

Second Experiment: We then performed the same experiment but compared it in a pairwise manner. Similarly to the previous experiment, we used 1 000 bootstrap samples.

Third Experiment: Since all graphs originate from the same monkey, there is a possibility that obtaining low p-values in the previous experiments is not a consequence of the difference between the distinct categories. To verify that the significance of the previous experiments was valid, we performed an ANOGVA test under the null hypothesis. In specific, we performed the following for each group G_i.

1. We split group G_i into two randomly sampled groups with no replacement, obtaining $G_{i,1}$ and $G_{i,2}$
2. We performed an ANOGVA test on these groups with 300 bootstrap samples.
3. We stored the calculated p-value.
4. We repeated this procedure 300 times, generating a distribution of p-values.

Suppose we explain low p-values because all graphs originate from the same monkey. In that case, performing ANOGVA using the described setup should give us mostly low p-values.

5.4   Results

First Experiment: For the first experiment, we obtained a p-value less than $\frac{1}{300}$. This shows that there is at least one sample of graphs that were generated differently.

Second Experiment: Table 3 shows the p-values obtained when comparing groups G_i and G_j. We note the low p-values, indicating that our method could differentiate between any two groups.

Third Experiments: Figure 2 shows the distribution of the p-values when comparing each group with itself. Any fear that previous low p-values might be because both groups originate from the same monkey can be eased by looking at the results of this experiment. We note a well-defined uniform distribution in each group, proving that the graphs from the same monkey are insufficient to justify a low p-value between groups.

Table 3. Results of second experiment:

	G_1	G_2	G_3	G_4	G_5
G_1		0.002	0	0	0
G_2	0.002		0	0	0.076
G_3	0	0		0	0
G_4	0	0	0		0
G_5	0	0.076	0	0	

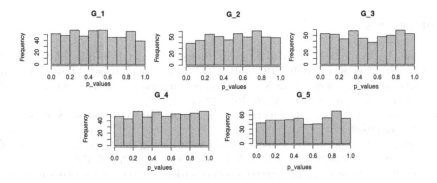

Fig. 2. Results of the third experiment.

Experiment Conclusion: We have shown that our methods can differentiate between the brain connectivity networks associated with all stages in the anesthesia experiment. These results promise that our methods can be used in future clinical trials.

6   Conclusion

To distinguish between populations of directed graphs, we explored measures based on the graph spectrum. We compared groups of graphs by calculating the Kullback-Leibler divergence between the graphs' spectra. This led to the development of ANOGVA, a non-parametric model for testing whether two or more groups of graphs share the same spectral distribution.

We demonstrated that our proposed method effectively distinguishes populations of graphs generated by different parameters, irrespective of the model used. Similarly, regular ANOVA on centrality measures can also distinguish various models. However, traditional ANOVA fails when centrality measures, like the number of edges in the Watts-Strogatz random model, remain unchanged. In our illustrative application with ECoG data, we successfully captured changes in the neural activity network of anesthetized monkeys. Unlike many classification methods, the proposed method can be used in clinical settings for diagnosing psychological conditions without the need for model training.

Our current approach is limited to single-edge graphs, in which a node i can only be connected to a node j via, at most, one edge. Multi-edge graphs, which permit several connections between two nodes, are not represented so simply by an adjacency matrix, and thus our method fails to apply.

Acknowledgements. This work has been supported by FAPESP grants 2018/21934-5, 2019/22845-9, and 2020/08343-8, CNPq grant 303855/2019-3 and 440245/2022-2, CAPES (finance code 001), Alexander von Humboldt Foundation, the Academy of Medical Sciences - Newton Fund, and Wellcome Leap.

References

1. Alon, U.: An Introduction to Systems Biology: Design Principles of Biological Circuits. Chapman & Hall/CRC Mathematical and Computational Biology (2006)
2. Barabasi, A.L., Oltvai, Z.N.: Network biology: understanding the cell's functional organization. Nat. Rev. Genet. **5**, 101–113 (2004)
3. Barabási, A.L., Albert, R.: Emergence of scaling in random networks. Science **286**, 509–512 (1999)
4. Bullmore, E., Sporns, O.: Complex brain networks: graph theoretical analysis of structural and functional systems. Nat. Rev. Neurosci. **10**, 186–198 (2009)
5. Csardi, G., Nepusz, T.: The igraph software package for complex network research. Interjournal Complex Syst. **1695** (2006)
6. Duong, T.: KS: kernel density estimation and kernel discriminant analysis for multivariate data in R. J. Stat. Softw. **21**(7), 1–16 (2007)
7. Erdős, P., Rényi, A.: On random graphs I. Publ. Math. Debrecen **6**, 290–297 (1959)
8. Freeman, L.: A set of measures of centrality based on betweenness. Sociometry **40**, 35–41 (1977)
9. Fujita, A., Vidal, M.C., Takahashi, D.Y.: A statistical method to distinguish functional brain networks. Front. Neurosci. **11**, 66 (2017)
10. Fujita, A., Silva Lira, E., De Siqueira Santos, S.: A semi-parametric statistical test to compare complex networks. J. Complex Netw. **8** (2020)
11. Fujita, A., Takahashi, D.Y., Balardin, J.B., Vidal, M.C., Sato, J.R.: Correlation between graphs with an application to brain network analysis. Comput. Stat. Data Anal. **109**, 76–92 (2017)
12. Lees-miller, J., et al.: Correlation between graphs with an application to brain network analysis. Comput. Stat. Data Anal. **109**, 76–92 (2017)
13. MacKay, D.J.: Information Theory, Inference, and Learning Algorithms, 1st edn. Cambridge University Press, Cambridge (2003)
14. Ramos, T.C., Mourão-Miranda, J., Fujita, A.: Spectral density-based clustering algorithms for complex networks. Front. Neurosci. **17**, 926321 (2023)
15. Ribeiro, A., Vidal, M., Sato, J., Fujita, A.: Granger causality among graphs and application to functional brain connectivity in autism spectrum disorder. Entropy **23**, 1204 (2021)
16. Sameshima, K., Baccala, L.: Methods in brain connectivity inference through multivariate time series analysis (2016)
17. Santos, S.S., Fujita, A.: statGraph: statistical methods for graphs (2017). www.cran.r-project.org/package=statGraph
18. Scott, J.: Social Network Analysis. Sage, Newcastle upon Tyne (2012)

19. Watts, D., Strogatz, S.: Collective dynamics of "small-world' networks. Nature **393**, 440–442 (1998)
20. Yanagawa, T., Chao, Z.C., Hasegawa, N., Fujii, N.: Large-scale information flow in conscious and unconscious states: an ECoG study in monkeys. PLoS ONE **8**(11), e80845 (2013)

Feature Selection Investigation in Machine Learning Docking Scoring Functions

Maurício Dorneles Caldeira Balboni[✉][ID], Oscar Emilio Arrua[ID],
Adriano V. Werhli[ID], and Karina dos Santos Machado[ID]

Computational Biology Laboratory - COMBI-Lab, Centro de Ciências
Computacionais, Universidade Federal do Rio Grande - FURG, Rio Grande, Brazil
mdcbalboni@gmail.com, karina.machado@furg.br

Abstract. The *in silico* evaluation of small molecules (ligands) and receptors (proteins) interactions is of great importance, especially in Drug Design. This is one of the principal computational methodologies that can be incorporated into the process of proposing new drugs, with the aim of reducing the high financial costs and time involved. In this context, molecular docking is a computer simulation procedure used to predict the best conformation and orientation of a ligand in the binding site of a target protein. These docking algorithms evaluate the protein-ligand complex interactions using scoring functions (SF). SF computationally quantify the complex binding affinity and can be divided into categories according to the methodology applied in their development: Physics-based, Empirical, Knowledge-based and Machine Learning. Machine Learning (ML) scoring functions train the SF considering features obtained from known protein-ligand complexes and experimental affinities. These SF rely heavily on the set of attributes that are used to train them. Thus, in this work, we use PCA, ANOVA and Random Forest to investigate how these feature selection methods impact the performance of three Machine Learning scoring functions trained with Support Vector Machines, Elastic Net Regularization and Neural Networks algorithms. The results show that Neural Networks can greatly benefit from Feature selection performed by Random Forests but not from ANOVA and PCA. The conclusions are that Feature selection can improve the results of regression and in this study Neural Networks combined with Random Forest is the best option.

Keywords: Rational Drug Design · Molecular Docking · Machine Learning · Feature Selection · Scoring Functions

1 Introduction

Living organisms carry the genetic code for producing a wide range of products, including proteins that play a crucial role in carrying out a myriad of biological

M. S. Reis and R. C. de Melo-Minardi (Eds.): BSB 2023, LNBI 13954, pp. 58–69, 2023.
https://doi.org/10.1007/978-3-031-42715-2_6

functions. Small molecules can interact with proteins altering their function making the study of these interactions highly significant in various research domains. Computational investigation of these interactions is continuously gaining relevance and finds extensive application in Drug Discovery [26]. Consequently, prior to conducting *in vitro* and *in vivo* experiments, researchers often perform *in silico* investigation as part of the Rational Drug Design (RDD) [15,21].

In RDD, docking simulations are an essential step in experiments. They are employed to predict the best conformation and estimate the binding affinity of a ligand (small molecule) in a binding site of a receptor (Protein) [20]. The Free Energy of Binding (FEB), or binding affinity, is computed by a scoring function [23,33]. The development of scoring functions (SF) is still a challenge [4] and different types have been developed. These SFs types are classified into four categories [18] related to the method that is used to obtain the protein-ligand binding affinity: (i) *Physics-based* that uses force fields to calculate protein-ligand binding; (ii) *Empirical* that calculates the protein-ligand fitness through the sum of individual terms that represent important energetic factor in protein-ligand binding; (iii) *Knowledge-based*, that obtains the score from summing pairwise statistical potentials between protein and ligand and (iv) *Machine Learning* methods that train scoring functions using features obtained from known protein-ligand binding experiments. Among these categories of scoring functions, the Machine Learning (ML) category is relatively new and promising as in general they present superior results [27,33].

Although ML scoring functions present very promising results, there are some aspects that require further investigation. There are many features (descriptors)[1] that can be obtained from a protein (e.g. frequency of amino acids in primary structure, secondary structure characteristics, number of atoms, chains, amino acids, etc.), from a ligand (2- and 3- descriptors like radius of gyration, number of atoms, rings, rotatable bonds, etc.) or from a protein-ligand complex (e.g. close contacts according to atom types, electrostatic interactions, hydrophobic contacts, etc.). Choosing which of these features are relevant for the training of ML methods is not a trivial task. Besides, the knowledge about which descriptors are more intimately related to binding affinity can shed some light in the discussion about protein-ligand interactions.

In this work we consider 723 molecular descriptors related to protein-ligand complexes. These descriptors are obtained using the following molecular feature extraction software: AA, DSSP, BINANA, PaDEL, RDDit 2D/3D, SASA and Vina. To select the most important features out of the 723 features available, three Feature Selection methods are used: (i) Analysis of Variance (**ANOVA**), (ii) Principal Component Analysis (**PCA**) and (iii) Random Forests (**RF**).

After selecting the most important features, the data sets composed of these features and the experimental pK_d are used to train three ML Regression methods: (i) Elastic net regularization (**ENR**), (ii) Neural Networks (**NN**) and (iii) Support Vector Machine (**SVM**) and obtain the values of binding affinity, $p\hat{K}_d$.

---

[1] In this paper, features and descriptors are treated as synonyms.

Using this experimental setting, we investigate how the Regression methods' performance is affected by Feature selection in the task of predicting binding affinity. Moreover, we explore the most important molecular descriptors found by the Feature selection methods to better understand how to improve the development of new scoring functions.

2   Material and Methods

Figure 1 shows our proposed methodology. The right panel presents a diagram of the steps that are executed in the methodology. Examples for each of these steps are presented in the left. All the steps were performed using in-house Python scripts and the Scikit-learn package [25].

First, the PDBbind 2018 Refined set [31] is retrieved. Various feature extraction tools (AA, DSSP, BINANA, PaDEL, RDKit 2D/3D, SASA and Vina) are employed to obtain the descriptors. After conducting exploratory data analysis (EDA) and preprocessing, we generate the input data. Next, we apply feature selection algorithms such as PCA, Anova, and RF to the input dataset. The resulting attributes are then used in Machine Learning regression methods, namely Elastic Net Regression, Neural Network, and SVM, to obtain the proposed ML scoring functions. Finally, the models are evaluated using two distinct metrics RMSE (Root Mean Square Error) and Correlation.

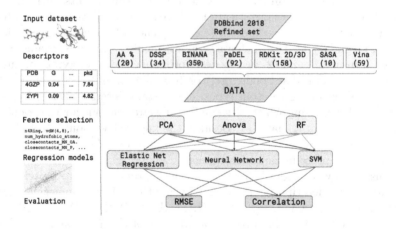

Fig. 1. Proposed methodology.

2.1   Protein-Ligand Complexes for Input Dataset

PDBBind-CN [31] is a comprehensive assembly of experimentally measured binding affinity data for complexes (protein-ligand, protein-protein, protein-nucleic acid, and nucleic acid-ligand) stored in Protein Data Bank (PDB). In this paper we are considering the PDBbind Refined set 2018 [19]. The refined set selects the protein-ligand complexes with better quality from the PDBBind General set using filters like binding data, crystal structures quality, and nature of the complexes [17]. The data package of PDBbind 2018 includes index files with information about the processed structures, receptor PDB file, ligands in Mol2, and SDF format and also files with only the protein pocket. PDBbind refined set 2018 has 4,463 complexes in total, from where we separate the core set (CASF 2016 [27], to use in future validation) that has 285 structures, totaling 4,178 complexes. A small part of these complexes presented errors during the generation of the features and was not considered. Thus, our input dataset based on PDBbind refined set 2018 has 4,152 complexes in total.

There are different experimental ways of expressing the protein-ligand binding affinity. In PDBbind 2018 the available experimental values of binding affinity are the dissociation constant (K_d), the inhibition constant (K_i), and the concentration at 50% inhibition(IC50) [19]. In this work we use the K_d because it is the most commonly employed to provide the binding affinity characterization. Values of K_d in PDBBind Refined Set varies according to the experimental method hence we use the normalized k_d. Taking into consideration the details presented above, the target attribute in our input files is the pK_d that corresponds to $-\log_{10} K_d$.

2.2   Feature Extraction

Starting from the PDBbind data, a total of 723 features are obtained from different sources. These features were extracted/obtained as follows, where the number in parentheses corresponds to the total of attributes of a given type followed by the source of the features (Receptor, Ligand or the Complexes):

- **AA % (20 - Receptor)**: these attributes correspond to the percentage of each amino acid type in the primary receptor structure. We calculated these using Biopython [3];
- **DSSP (34 - Receptor)**: we considered 34 attributes selected from the list used by [14]. This list is based on protein secondary structure functional characteristics calculated by DSSP [12]. These attributes are: the total number of residues, chains, SS-bridges, SS-bridges intra-chain and inter-chain, hydrogen bonds in anti-parallel bridges and parallel bridges, hydrogen bonds of different types (O(I) H-N(I-5), O(I) H-N(I-4) and so on).
- **BINANA (350 - Receptor - Ligand - Complex)**: The *Python* implemented algorithm BINANA (BINding ANAlyzer) [6] calculates descriptors for characterizing protein-ligand binding. These attributes were considered in the scoring function NNSCORE 2.0 [5] and are divided into:
  1. **close contacts** corresponds to the frequency of all ligand and protein atoms types that are within 2.5 Å or 4.0 Å of each other;
  2. **electrostatic interactions** calculates, for each atom-type pair of atoms within 4.0 Å of each other, the sum of electrostatic energy using Gasteriger partial charges [22];

3. **binding-pocket flexibility** starts verifying for each receptor atom within 4.0 Å of any atom of the ligand if it belongs to a side chain or backbone, then check if this receptor atom is in a α-sidechain, α-backbone, β-sidechain, β-backbone, other-sidechain, other-backbone and finally counts these frequencies;

4. **hydrophobic contacts** simply verify the number of times a ligand carbon atom is within 4.0 Å of a receptor carbon atom and calculates the frequency where this receptor carbon atom is according to side-chain/backbone/secondary structure;

5. **hydrogen bonds** counts the number of hydrogen bonds according to the side-chain/backbone/secondary structure of the receptor atom and where is the hydrogen-bond donor (ligand or receptor) generating twelve attributes;

6. **salt bridges** corresponds to the possible salt bridges between the receptor and the ligand categorized by the secondary structure of the protein residue α-helix, β-sheet or other;

7. **π interactions** starts identifying and characterizing all aromatic rings of receptor residues and five or six aromatic rings, aromatic or not of ligands followed by the identification of π-π stacking, T-stacking, and cation-π interactions between these rings according to the secondary structure of the receptor residue containing the associated aromatic ring or charged functional group α-helix, β-sheet, or other;

8. **ligand information** calculates the number of atoms of each atom type and the number of rotatable bonds of the ligands (according to Autodock);

- **PaDEL (92 - Ligand):** The software PaDEL calculates 1-, 2- and 3D ligands descriptors and 10 types of fingerprints [34]. From these, we chose to consider 92 2D descriptors that comprehend:

  1. **atom count descriptors:** total number of heavy atoms, H, B, C, N, O, S, P, F, Cl, Br, I, and halogens;

  2. **bound count descriptors:** the total number of single and/or double bonds considering and not considering hydrogens and the total number of triple or quadruple bonds;

  3. **rings count descriptors:** number of single and Ring count descriptors (total number of rings and/or fused rings, number of 3-, 4-, 5-, 6-, 7-, 8-, 9-, 10-, 11-, 12-, > 12-membered rings and/or fused rings, number of rings and/or fused rings containing heteroatoms, number of 3-, 4-, 5-, 6-, 7-, 8-, 9-, 10-, 11-, 12-, >12-membered rings and/or fused rings with heteroatoms).

- **RDKit(158 - Ligand):** The open-source toolkit for cheminformatics RDKit generates 2-,3D descriptors about ligands [16]. In our proposed method, 147 2D descriptors like the number of aliphatic rings; saturated/aliphatic heterocycles; some polarity counts; MOE-type descriptors, and so on;

- **SASA (10 - Receptor):** These descriptors are related to solvent-accessible surface area (SASA). For details, see material of [30].

– **Vina (59 - Ligand - Complex):** 58 terms implemented by AutoDock Vina source code [7, 29]: protein-ligand **interactions terms** (gauss, repulsion, van der walls, electrostatic, hydrogen bond, hydrophobic, non-hydrophobic, autodock 4 solvatation) and **ligand dependent counts** (number of torsions, rotors, heavy and hydrophobic atoms, hydrogen bonds, ligand length). On this group of attributes, we are also considering the **vina score** [7, 29].

2.3 Preprocessing

Before applying feature selection, we first performed a Exploratory Data Analysis (EDA). After the EDA, we removed attributes where the entropy was equal to zero as they do not offer information gain for generating the models. In addition to this, we normalized all the numeric attributes to have values between 0 and 1. This resulted in an input data set with a dimension of 4,152 (instances) × 502 (attributes).

2.4 Feature Selection

Dimensionality reduction is a process that aims to reduce the dimensionality of the training set, in order to obtain a set of features that are more relevant to training the model. This process aims to reduce overfitting, increase the accuracy of the results and reduce training time [8]. Based on this, we propose to investigate the impact of distinct feature selection methods in the prediction of pK_d. Three different feature selection methods were exploited:

– **PCA:** Principal Component Analysis [24] is an statistical multivariate technique that transforms the original data set in a new data set of principal components. Each feature in the new data set is a linear combination of all original features, maintaining the maximum of information in terms of data variability. To decide how many features have to be kept to maintain 90% of the data variability, Fig. 2 shows the Elbow Plot where it is possible to verify that 77 features are necessary.

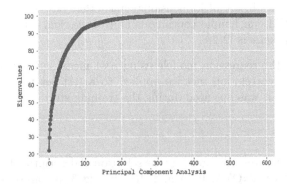

Fig. 2. Elbow Plot. This plot shows in the horizontal axis the number o components and in the vertical axis the Eigenvalues, or the percentage of variability that is kept. To explain 90% of the data variability 77 features are necessary.

- **Anova:** The statistic for Analyses of Variance (Anova), or the F-statistic, is calculated and used to decide if there are significant differences between the means of multiple features [13]. This allows the ordering of features according to their degree of importance, following the F-value. Thus, in order to compare the feature selection methods, we pick the first 77 features considering the Anova ranking of features.
- **Random Forest (RF):** A RF is composed by a large number of individual trees, operating as an ensemble classifier [1]. It builds unpruned individual trees using bootstrap samples of the data. At each split, a random subset of variables is considered as the candidate set of variables. Random Forest algorithm uses both bagging and random variable selection for tree building. Besides, a RF gives the order of importance of the attributes [11]. In this work, this order of importance is used for selecting the first 77 most important features.

2.5   Regression Methods and Validation

Three regression methods are applied to infer the pK_d using the selected features, they are:

- **Elastic Net Regularization (ENR):** is a weighted combination of the penalties of Ridge Regression (L2 regularization) and Lasso (L1 regularization) [10]. ENR reduces the problems of Lasso by adding to its penalty a quadratic term that when used alone is known as ridge regression;
- **Neural Networks (NN):** NNs is an artificial intelligence algorithm inspired by the human brain that can be applied to various learning tasks, including classification problems. They are very flexible models, trained with examples for which the correct classification is known. A typical NN is composed of a topology and its parameters. The topology defines the number of layers (an input layer, one or more hidden layers, and an output layer) and the number of nodes in each layer. The parameters for such topology are learned from the data during the training [32]. In this work, we employed a NN with two layers of 50 neurons, and the threshold was set to 0.01.
- **Support Vector Machines (SVM):** are classifiers that learn from the input data hyper-planes in a multidimensional space that best separate the labeled classes [9]. SVM can handle continuous and categorical target attributes by using different kernels [2,28].

We evaluate the regression models using root mean squared error (RMSE) and the Pearson correlation metrics applied to a test set that is part of the original dataset but was not used during the training step.

3   Results

First, we present the results regarding the selection of descriptors by each Feature extraction method: ANOVA, PCA and, RF. In Fig. 3 we can observe the total number of descriptors of each type selected by each method.

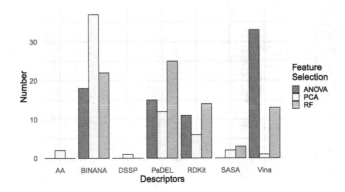

Fig. 3. Total number of descriptors of each type selected by the algorithms ANOVA, RF and PCA.

Considering the 77 features selected by ANOVA, we can notice that 18 are from BINANA, 15 from PaDEL, 11 from RDKit and 33 from Vina. This feature selection method did not select any descriptors of SASA, AA or DSSP types. The 61 descriptors selected by PCA vary between all types: 2 of AA, 1 of DSSP, 37 of BINANA, 12 of PaDEL, 6 of RDKit, 1 of Vina and 2 of SASA. Finally, considering the RF as the feature selection method, the most frequent descriptor type is PaDEL (25) followed by BINANA (22), RDkit (14), Vina (13) and, SASA (3). Only descriptors of AA and DSSP types were not chosen by this method.

Figure 4 details the selected features according to each method. We can observe that we had 188 different features and only 2 descriptors were selected by all methods ($cc\_MN.OA\_2.5$ and $cc\_A.F\_4$). These descriptors are from BINANA and correspond to close contacts between the atom types MN-OA in a distance of 2.5 Å and between A and F in a distance of 4 Å (the definition about atom types are based on AutoDock [29]). From these 188 descriptors, 24 were chosen by at least two methods, from which 14 are also of type close contacts and 4 are related to electrostatic contacts from BINANA. These results indicate that information about the frequency of contacts between protein and ligand, at distances of 2.5 Å and 4 Å, are interesting descriptors to be considered in the future.

The results in regard to the ML methods are summarised in Fig. 5. To compare the results obtained with different Regression methods: (i) Elastic net regularization (**ENR**), (ii) Neural Networks (**NN**) and (iii) Support Vector Machine (**SVM**) and Feature Selection methods: (i) Analysis of Variance (**ANOVA**), (ii) Principal Component Analysis (**PCA**) and (iii) Random Forests (**RF**) two metrics are used, Pearson Correlation and RMSE.

In Fig. 5, Regression and Feature Selection methods are presented respectively in vertical and horizontal axis. The Selection method **ALL** is the result of considering all the 502 non-zero features. The left and right panel present respectively the Pearson correlation and RMSE metrics. The values of both metrics is shown inside the squares. The color in each square follows a gradient that varies from worst result (clearer color) to best result (darker color). The better results

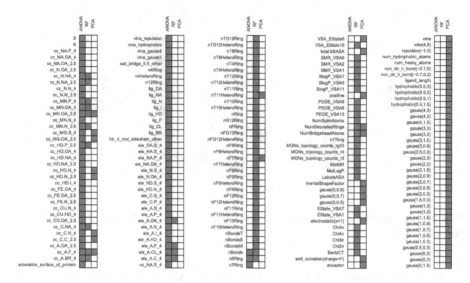

Fig. 4. Selected Features. The grid shows in rows the name of the Features and in columns the Feature selection algorithms. The graph is divided in 5 panels due to the number o Features, each panel is a continuation of the previous panel. Dark coloured squares indicate the Features that were selected by each method.

are closer to 1 (one) for Pearson correlation and are closer to 0 (zero) for RMSE, hence the inverted color gradients. The gray squares are relative to values that are not available.

4   Discussion and Future Work

Scoring functions for protein-ligand molecular docking based on ML are very promising. However, further feature selection studies are needed to improve their performance. In this work we have explored how Feature selection methods impact different Regression models. The number of Features to be considered, 77, was obtained from the PCA, by analysing the elbow curve, Fig. 2. This number of 77 features is used in the other methods. Moreover, the results of using all the features are also presented.

In relation to the descriptors selected by ANOVA, it is important to mention that 33 out of the 77 are of vina type, which means that, according to this feature selection method, this descriptor type is important and related to the pkd binding affinities. The only method that selected vina score was PCA.

Concerning the selected features by all methods, we can observe that they vary and did not converge to the same subset of descriptors. Considering all the descriptors, we have a total of 188 of which only 2 are common to ANOVA, RF and PCA and 24 more are present in at least two of them. Regarding the descriptors types, we can conclude that descriptors types related only to the receptor (AA and DSSP) were rarely chosen. On the other hand, the descriptors

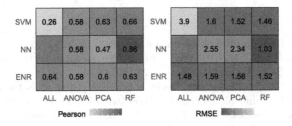

Fig. 5. Results of different Regression methods and Feature Selection methods evaluated with two different metrics: Pearson Correlation (left) and RMSE (right). In the vertical axis, Regression methods are: Elastic net regularization (ENR), Neural Networks (NN) and, Support Vector Machine (SVM). In the horizontal axis, Feature Selection methods are: ALL where all features are considered, Analysis of Variance (ANOVA), Principal Component Analysis (PCA) and Random Forests (RF). The color gradient varies from worst results (clearer blue) to best results (darker blue). The gray color indicates that results are not available. (Color figure online)

more frequent on the 188 list (Fig. 4 correspond to the frequency of close contacts, for example, the 14 descriptors from BINANA selected by at least two methods.

With respect to the ML results, for NN with ALL the features the simulations did not converge, hence there are not results for this setting. The results show that SVM with ALL is the worst result and NN with RF is the best result. All other settings are very similar, however, it is possible to see that ENR with no feature selection (ALL) performed slightly better than with feature selection. This is not surprising giving that ENR has its Lasso component that is responsible by feature selection, i.e., ENR has an embedded feature selection.

The main conclusions are that Feature selection is important in this problem and has the potential to improve the results of ML methods. However, the impact of Feature selection is different for each regression algorithm and they should be applied with caution. Most importantly, this study shows that the combination of methods and features can be used to develop a new protein-ligand docking scoring function based on ML.

As future work we propose enhancing the EAD and preprocessing steps by considering the prerequisites of each FS method. Additionally, we plan to evaluate alternative Machine Learning algorithms to generate the scoring functions. Moreover, we aim to compare the performance of the proposed scoring function with other SFs using bechmarks such as CASF 2016 [27].

Acknowledgments. The authors acknowledge the financial support given by CAPES Financial Code 001, CNPq grants 439582/2018-0 and 440363/2022-5, and FAPERGS processes 22/2551-0000385-0 and 22/2551-0000390-7.

References

1. Breiman, L.: Random forests. Mach. Learn. **45**(1), 5–32 (2001)
2. Chang, C.C., Lin, C.J.: LIBSVM: a library for support vector machines. ACM Trans. Intell. Syst. Technol. **2**, 27:1–27:27 (2011)
3. Cock, P.J.A., et al.: Biopython: freely available Python tools for computational molecular biology and bioinformatics. Bioinformatics **25**(11), 1422–1423 (2009)
4. Crampon, K., Giorkallos, A., Deldossi, M., Baud, S., Steffenel, L.A.: Machine-learning methods for ligand-protein molecular docking. Drug Discovery Today **27**(1), 151–164 (2022)
5. Durrant, J.D., McCammon, J.A.: NNScore: a neural-network-based scoring function for the characterization of protein-ligand complexes. J. Chem. Inf. Model. **50**(10), 1865–1871 (2010)
6. Durrant, J.D., McCammon, J.A.: BINANA: a novel algorithm for ligand-binding characterization. J. Mol. Graph. Model. **29**(6), 888–893 (2011)
7. Eberhardt, J., Santos-Martins, D., Tillack, A.F., Forli, S.: AutoDock vina 1.2. 0: new docking methods, expanded force field, and python bindings. J. Chem. Inf. Model. **61**(8), 3891–3898 (2021)
8. Guyon, I., Elisseeff, A.: An introduction to variable and feature selection. J. Mach. Learn. Res. **3**, 1157–1182 (2003)
9. Han, J., Pei, J., Tong, H.: Data Mining: Concepts and Techniques. Morgan Kaufmann (2022)
10. Hans, C.: Elastic net regression modeling with the orthant normal prior. J. Am. Stat. Assoc. **106**(496), 1383–1393 (2011)
11. Ishwaran, H., Lu, M.: Standard errors and confidence intervals for variable importance in random forest regression, classification, and survival. Stat. Med. **38**(4), 558–582 (2019)
12. Kabsch, W., Sander, C.: Dictionary of protein secondary structure: pattern recognition of hydrogen-bonded and geometrical features. Biopolymers **22**(12), 2577–2637 (1983). https://doi.org/10.1002/bip.360221211. www.onlinelibrary.wiley.com/doi/abs/10.1002/bip.360221211
13. Kumar, M., Rath, N.K., Swain, A., Rath, S.K.: Feature selection and classification of microarray data using MapReduce based ANOVA and K-nearest neighbor. Procedia Comput. Sci. **54**, 301–310 (2015)
14. Kundu, I., Paul, G., Banerjee, R.: A machine learning approach towards the prediction of protein-ligand binding affinity based on fundamental molecular properties. RSC Adv. **8**(22), 12127–12137 (2018)
15. Kuntz, I.D.: Structure-based strategies for drug design and discovery. Science **257**(5073), 1078–1082 (1992)
16. Landrum, G.: RDKit documentation. Release **1**(1–79), 4 (2013)
17. Li, Y., et al.: Comparative assessment of scoring functions on an updated benchmark: 1. Compilation of the test set. J. Chem. Inf. Model. **54**(6), 1700–1716 (2014)
18. Liu, J., Wang, R.: Classification of current scoring functions. J. Chem. Inf. Model. **55**(3), 475–482 (2015)
19. Liu, Z., et al.: Forging the basis for developing protein-ligand interaction scoring functions. Acc. Chem. Res. **50**(2), 302–309 (2017)
20. Lybrand, T.P.: Ligand-protein docking and rational drug design. Curr. Opin. Struct. Biol. **5**(2), 224–228 (1995)
21. Mahapatra, M.K., Karuppasamy, M.: Fundamental considerations in drug design. In: Computer Aided Drug Design (CADD): From Ligand-Based Methods to Structure-Based Approaches, pp. 17–55. Elsevier (2022)

22. Morris, G.M., et al.: AutoDock4 and AutoDockTools4: automated docking with selective receptor flexibility. J. Comput. Chem. **30**(16), 2785–2791 (2009)

23. Onodera, K., Satou, K., Hirota, H.: Evaluations of molecular docking programs for virtual screening. J. Chem. Inf. Model. **47**(4), 1609–1618 (2007)

24. Pearson, K.: Principal components analysis. London Edinburgh Dublin Philosophical Mag. J. Sci. **6**(2), 559 (1901)

25. Pedregosa, F., et al.: scikit-learn: machine learning in Python. J. Mach. Learn. Res. **12**, 2825–2830 (2011)

26. Piñero, J., Furlong, L.I., Sanz, F.: In silico models in drug development: where we are. Curr. Opin. Pharmacol. **42**, 111–121 (2018)

27. Su, M., et al.: Comparative assessment of scoring functions: the CASF-2016 update. J. Chem. Inf. Model. **59**(2), 895–913 (2018)

28. Tan, P.N., Steinbach, M., Kumar, V.: Introduction to Data Mining. Pearson (2016)

29. Trott, O., Olson, A.J.: AutoDock vina: improving the speed and accuracy of docking with a new scoring function, efficient optimization, and multithreading. J. Comput. Chem. **31**(2), 455–461 (2010)

30. Wang, C., Zhang, Y.: Improving scoring-docking-screening powers of protein-ligand scoring functions using random forest. J. Comput. Chem. **38**(3), 169–177 (2017)

31. Wang, R., Fang, X., Lu, Y., Wang, S.: The PDBbind database: collection of binding affinities for protein-ligand complexes with known three-dimensional structures. J. Med. Chem. **47**(12), 2977–2980 (2004)

32. Wang, S.C.: Artificial neural network. In: Interdisciplinary Computing in Java Programming, pp. 81–100. Springer, Boston (2003). https://doi.org/10.1007/978-1-4615-0377-4_5

33. Yang, C., Chen, E.A., Zhang, Y.: Protein-ligand docking in the machine-learning era. Molecules **27**(14), 4568 (2022)

34. Yap, C.W.: PaDEL-descriptor: an open source software to calculate molecular descriptors and fingerprints. J. Comput. Chem. **32**(7), 1466–1474 (2011)

Using Natural Language Processing for Context Identification in COVID-19 Literature

Frederico Carvalho[1,2]([✉]) [iD], Diego Mariano[1,2] [iD], Marcos Bomfim[2] [iD],
Giovana Fiorini[1] [iD], Luana Bastos[1] [iD], Ana Paula Abreu[1] [iD], Vivian Paixão[1] [iD],
Lucas Santos[1] [iD], Juliana Silva[1] [iD], Angie Puelles[1] [iD], Alessandra Silva[1] [iD],
and Raquel Cardoso de Melo-Minardi[1] [iD]

[1] Laboratory of Bioinformatics and Systems, Institute of Exact Sciences, Department of
Computer Science, Universidade Federal de Minas Gerais, Belo Horizonte, Brazil
fcc073@gmail.com
[2] SimplificaSUS Developers Team, Brasília, Brazil

Abstract. The COVID-19 pandemic led to an unprecedented volume of articles published in scientific journals with possible strategies and technologies to contain the disease. Academic papers summarize the main findings of scientific research, which are vital for decision-making, especially regarding health data. However, due to the technical language used in this type of manuscript, its understanding becomes complex for professionals who do not have a greater affinity with scientific research. Thus, building strategies that improve communication between health professionals and academics is essential. In this paper, we show a semi-automated approach to analyze the scientific literature through natural language processing using as a basis the results collected by the "Scientific Evidence Panel on Pharmacological Treatment and Vaccines – COVID-19" proposed by the Brazilian Ministry of Health. After manual curation, we obtained an accuracy of 0.64, precision of 0.74, recall of 0.70, and F1 score of 0.72 for the analysis of the using-context of technologies, such as treatments or medicines (i.e., we evaluated if the keyword was used in a positive or negative context). Our results demonstrate how machine learning and natural language processing techniques can greatly help understand data from the literature, taking into account the context. Additionally, we present a proposal for a scientific panel called SimplificaSUS, which includes evidence taken from scientific articles evaluated through machine learning and natural language processing methods.

Keywords: COVID-19 · SimplificaSUS · Literature · NLP

1 Introduction

In 2020, the COVID-19 pandemic spread worldwide, causing irreparable human losses [1–4]. COVID-19 (CoronaVirus Disease 2019) is a respiratory disease caused by the SARS-CoV-2 (Severe Acute Respiratory Syndrome Coronavirus 2), a single-stranded sarbecovirus with a positive-sense RNA [5]. Thus, the scientific community responded with an unprecedented amount of studies seeking to understand the origins, mechanisms of action, and possible treatments for the disease [6–8].

© The Author(s), under exclusive license to Springer Nature Switzerland AG 2023
M. S. Reis and R. C. de Melo-Minardi (Eds.): BSB 2023, LNBI 13954, pp. 70–81, 2023.
https://doi.org/10.1007/978-3-031-42715-2_7

Governments and health organizations aimed to provide the means to help the results of new studies reach physicians, health professionals, and society in general. For instance, the Brazilian Ministry of Health team proposed a panel to summarize the COVID-19 scientific literature available until that moment and show the results using user-friendly data visualizations [9]. The "Scientific Evidence Panel on Pharmacological Treatment and Vaccines - COVID-19" aimed to gather real-time information on technical-scientific publications from indexed and pre-printed journals that investigate the efficacy, safety, and effectiveness of drugs and biological products used for the treatment and prevention of COVID-19. Until the evaluated date (06/17/2022), the panel summarized information from 2,147 manually curated scientific articles. However, the large amount of data can be detrimental to interpreting the problem since the vast amount can lead to extremely different understandings of the information, making it difficult to transmit the knowledge.

For example, in one of the analyzes proposed by the panel, they raised more than 500 technologies used in the treatment mentioned in the articles. These technologies could be vaccines, therapies, drugs, among others. However, the proposed visualizations in their panel considered only the count of times a technology was mentioned. As the context was not considered, this could lead to misinterpretations. For example, the most cited technology in articles published up to that point was hydroxychloroquine, a drug used to treat malaria. Several studies pointed to the possibility of using hydroxychloroquine to combat COVID-19, in a strategy known as drug repositioning [10]. However, later studies and literature reviews showed that the efficiency of hydroxychloroquine in combating COVID-19 could not be proven [11–15].

In fact, the analysis of some of the articles cited in the panel already showed that both chloroquine and hydroxychloroquine did not have proven effectiveness [16]. However, considering only the number of times these technologies are cited in the literature may give a false impression that their use was effective. Additionally, a manual analysis of each technology and the context of mention in each article in real-time would require an unviable number of dedicated personnel.

We hypothesize that text mining and natural language processing (NLP) techniques could be used to identify the context in which technology is mentioned in the scientific literature. One such technique is sentiment analysis, often used to analyze whether users' reviews about a certain product are positive or negative. In this context, this technique has the potential to promptly identify which papers cite each technology as effective or non-effective for a given disease, which could contribute to the faster adoption of more effective public policies by the health authorities.

Here we show the results of our NLP-based analyzes of the data presented in the "Scientific Evidence Panel on Pharmacological Treatment and Vaccines - COVID-19". Our tool aims to collect, analyze, and present evidence taken from scientific articles evaluated through machine learning techniques and natural language processing. To evaluate our results, our team manually analyzed the citation context of each technology in the papers from the panel. Then, we performed a comparison with a panel produced by the Brazilian Ministry of Health.

2   Material and Methods

2.1   Data Collection

We collected metadata from 2,147 articles from the Scientific Evidence Panel on Pharmacological Treatment and Vaccines - CoViD-19, available at: https://infoms.saude.gov. br/extensions/evidencias_covid/evidencias_covid.html. Titles, abstracts, authors, DOI, journal name, and ISSN were obtained from PubMed API (https://www.ncbi.nlm. nih.gov/home/develop/api/) using in-house Python scripts. Qualis strata were collected from Sucupira (https://sucupira.capes.gov.br/sucupira/public/consultas/coleta/vei culoPublicacaoQualis/listaConsultaGeralPeriodicos.jsf) and journal impact factor values were obtained from SJC - Scimago Journal & Country Rank (https://www.scimag ojr.com/journalrank.php). The list of technologies was also obtained from the Scientific Evidence Panel. Details were included in the Supplementary Material. We also used Drug Central's database of FDA, EMA, PMDA Approved Drugs (https://drugcentral. org/download) to identify other technologies and treatments not considered in the initial database.

2.2   NLP Analyses

In natural language processing, sentiment analysis models usually perform the task of analyzing if a given text is referring to a certain product or technology in a positive or negative light. These models are commonly used in industries such as social media monitoring, customer service and market research, where many customer-generated texts (reviews) need to be analyzed to identify public mood and help inform strategic decision-making.

Considering the specific context of evaluating whether medications are referred to in the articles as effective, our first step was to use regular expressions search to identify which medications or therapies listed in the Drug Central's FDA, EMA, and PMDA Approved Drugs database were mentioned in each paper. This approach is preferable to named entity recognition techniques since we are searching specifically for medications and therapies, not general entities.

Since not every paper is open-access, we only evaluated the titles and abstracts. After identifying the technologies cited in these sections, we separated the sentence in which they were cited as well as the next sentence and applied the Valence Aware Dictionary and sEntiment Reasoner (VADER) model to evaluate the sentiment of the two sentences [17]. If the technology was cited in more than one sentence, the procedure was repeated for each pair. Then, the average score was computed to indicate if the overall sentiment related to that technology was positive or negative.

Considering the aforementioned procedure, papers without abstracts or not in English were discarded before the processing. The overall results were then recorded in a CSV file, indicating which technologies were positively or negatively cited in each paper.

2.3 Manual Curation and Data Evaluation Metrics

Through a meticulous process of manual curation, each paper in the database received a label. Our team determined whether or not each study mentioned the highlighted treatment or technology as effective or not. Papers mentioning multiple technologies had multiple rows to allow for separate evaluation of each technology. On the other hand, papers that did not mention any technology or treatment present in Drug Central's database were eliminated. This removal is in accordance with the behavior of the proposed method, which uses regular expression search to determine which therapies are mentioned by the paper and eliminates from analysis those that do not mention any valid treatment, as mentioned in the previous section.

To avoid biases in the analysis, the evaluators were not aware of the label assigned to each paper by the sentiment analysis model. The final result was a manually curated database that was used as the "gold standard" for assessing the accuracy, precision, recall, and F1-score of the sentiment analysis model by comparing manual and model classifications.

3 Results and Discussion

In this study, we evaluated technologies cited in articles, reviews, comments, and pre-prints published during the COVID-19 pandemic (Fig. 1A). We propose that using natural language processing approaches could benefit rapid analysis and help make more effective public policy decisions. Our main objective in this work is to allow a semi-automatic analysis of datasets obtained from the scientific literature to obtain an initial overview. This analysis should be semi-automatic as a completely automatic analysis could include biases that would impair our understanding. Furthermore, a fully manual analysis would be time-consuming.

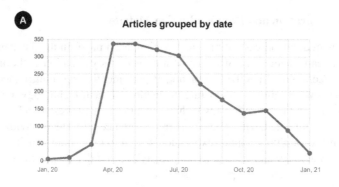

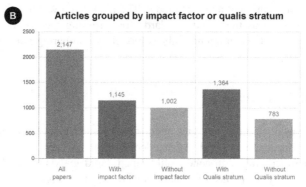

Fig. 1. Data overview. (A) Articles published by date. (B) Articles grouped by impact factor or Qualis stratum.

Our initial analysis aimed to understand the importance of the means used to publish these manuscripts. For this, we use two different metrics: (1) the impact factor and (2) the Qualis stratum. The impact factor is a metric that considers the average citation of a journal in recent years. The Qualis stratum is a metric used by Brazilian agencies to assess the quality of journals in which Brazilian researchers have already published. Journals without an assigned impact factor or Qualis stratum tend to be less recognized by the scientific community. Therefore, this is an initial factor in the quality of the published manuscript.

From 2,147 manuscripts, we detected that almost half of them do not have an impact factor or Qualis stratum (Fig. 1B).

3.1 Technology Mentions

The "Scientific Evidence Panel on Pharmacological Treatment and Vaccines - COVID-19" provides a list of technologies cited in the manuscripts. For each manuscript, the authors manually selected technologies used to treat COVID-19. Then, they evaluated the most cited technologies but did not evaluate the context in which the keyword was mentioned.

In the original panel, a total of 564 technologies were detected. However, when analyzing the list of technologies, we found many repeated or overlapping terms (in Portuguese) such as "Colchicine", "colchicine", and "anti-inflammatory drugs". In this initial form, the most cited technologies were: Hydroxychloroquine, tocilizumab, Lopinavir, Ritonavir, chloroquine, Azithromycin, remdesivir, corticosteroids, convalescent plasma, Angiotensin Converting Enzyme 2 Inhibitors (ACE 2), Angiotensin II Receptor Antagonists, prednisone, favipiravir, Umifenovir, heparin, Immunoglobulin, oseltamivir, Ivermectin, Ribavirin, Vitamin D, alpha interferon, Anakinra, cell therapy, interferon, darunavir, dexamethasone, BCG vaccine, vaccines, and so on.

Using the terms described in the Drug Central database, we identified 72 unique technologies mentioned in the titles and/or in the abstracts of the analyzed papers. Only 14 of the recognized therapies and medicines were cited in at least five different articles: hydroxychloroquine (90), tocilizumab (59), vaccines (35), azithromycin (28), chloroquine (25), remdesivir (22), anakinra (12), heparin (8), methylprednisolone (8), ribavirin (8), oxygen therapy (8), ivermectin (7), lopinavir (6) and famotidine (5). We evaluated the context using the sentiment analysis model and considered those whose context was classified as negative (negative sentiment score) as being referred to as non-efficient or prejudicial, while those described in a positive context (positive sentiment score) as efficient or beneficial for the COVID-19 treatment. The results were summarized as shown in Fig. 2.

Technology mention context analysis
by Natural Language Processing (NLP)

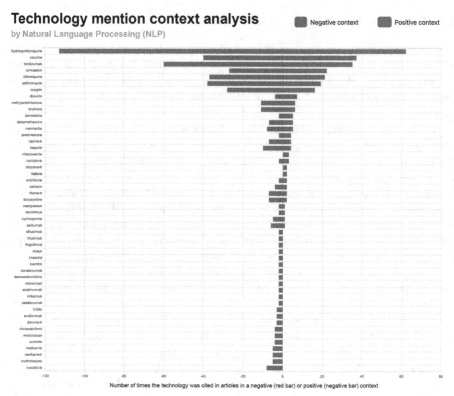

Fig. 2. Predictions of technology use depending on context. Red bars indicate quotes in a negative context. Blue bars indicate citations in a positive context.

3.2 Sentiment Analysis Model Performance Evaluation

The manually curated gold standard database consisted of 410 scientific articles that mentioned treatments and/or therapies for COVID-19. A total of 263 (~64%) of these papers highlighted the mentioned technologies as effective while 147 (~36%) of them cited the technologies as non-effective.

By comparing the results predicted by the NLP model with those obtained by manual curation, we obtained the results summarized in Table 1.

The comparison of the manual curation with the prediction indicates that the model's predictions were 64.88% accurate. The precision and recall achieved were 73.71% and 70.34%, respectively. These results suggest that the model has reasonable accuracy and capacity to identify a particular context in which the technologies were cited in the literature.

Table 1. Results of NLP analysis.

Metric	Value
Accuracy	0.6488
Precision	0.7371
Recall	0.7034
F1 score	0.7198

Table 2 illustrates some examples of predictions and the manually curated attributions for each statement.

Table 2. Eight classification examples: three negative and three positive correctly predicted, one negative and one positive mispredicted. Columns "prediction" and "real": negative context (0) and positive context (1). The complete table is available in the Supplementary material (https://github.com/LBS-UFMG/SimplificaSUS).

#	Technology	Prediction	Real	Statement evaluated	Source
1	azithromycin	0	0	"Hydroxychloroquine, chloroquine, and azithromycin produced no clinical evidence of efficacy in randomized controlled clinical trials (RCT)."	[18]
2	hydroxychloroquine	0	0	"Our study did not support the use of hydroxychloroquine plus atazanavir/ritonavir in patients who had SpO2 < 90% at the time of hospital admission."	[19]
3	fingolimod	0	0	"Our case suggests that discontinuation of fingolimod during COVID-19 could imply a worsening of SARS-CoV2 infection."	[20]
4	vaccine	1	1	"Therefore, development of a safe and effective vaccine against COVID-19 is an urgent public health priority."	[21]
5	anakinra	1	1	"Based on what we experienced in this case, anakinra could be an effective and reliable option in COVID-19-associated pericarditis"	[22]
6	remdesivir	1	1	"[…] to reconcile results to determine patient populations that may optimally benefit from remdesivir therapy"	[23]

(continued)

Table 2. (*continued*)

#	Technology	Prediction	Real	Statement evaluated	Source
7	chloroquine	1	0	"[…] the chloroquine hype, fueled by low-quality studies and media announcements, has yielded to the implementation of more than 150 studies worldwide."	[24]
8	ruxolitinib	0	1	"Rux treatment for COVID-19 in patients with hyperinflammation is shown to be safe with signals of efficacy in this pilot case series for CRS-intervention to prevent or overcome multiorgan failure."	[25]

For instance, in line 1 of Table 2, we show the results for the context analysis for the technology "azithromycin" in the article "Systematic review on the therapeutic options for COVID-19: clinical evidence of drug efficacy and implications", published by Abubakar et al. in 2020 in the Infection and Drug Resistance journal [18]. Our NLP analysis predicted that the word was used in a negative context (0). Indeed, a reviewer also manually attributed that this keyword was used in a negative context (0). We could verify this by analyzing the sentence in which it was cited: "[…] azithromycin produced no clinical evidence of efficacy […]". Hence, we can conclude that the NLP analysis correctly predicted the context of mention.

In [19], the authors affirm that their study "did not support the use of hydroxychloroquine […]". In this case, both our method and the reviewer attributed a negative context for the keyword "hydroxychloroquine", predicting correctly once again. Also, in [20], our method again correctly predicted the using context for the "fingolimod" keyword.

In lines 4, 5, and 6 of Table 2, we can see the predictions for the using context for the words "vaccine", "anakinra", and "remdesivir", in the papers [21, 22], and [23], respectively. In these cases, our method predicted a positive context, which was proved by manual curation.

The table also shows two failure samples, highlighted in lines 7 and 8. The analysis of these cases shows that the model sometimes fails to correctly classify if a treatment is effective or not in situations where it is described in sentences that also refer to the disease symptoms or external situations. In case 7, the prevalence of words and expressions usually seen in positive reviews (*e.g.*, "hype", "media announcements", "more than 150") might be the case for the incorrect classification as positive. Meanwhile, the description in case 8 has many words usually related to negative contexts but that is not related to the medication, such as "hyperinflammation", "intervention", "prevent", "overcome" and "failure", which may explain why the model attributed a negative context to the medication.

The examples above show that, despite achieving an acceptable performance in classifying which technologies and therapies are described mostly as effective or not effective in the literature, the model still fails in several common cases where the description of context and the disease symptoms are given close to the name of the medication. This suggests that a pre-trained model might not be the best option for the task selected, and building a more robust model trained on a larger database specifically designed for this use case might result in better performance for this task. We have future prospects to improve the model presented by testing it with other case studies.

SimplificaSUS Panel

As a secondary objective, we also present the SimplificaSUS panel. Our objective, in this case, is to provide a user-friendly tool that facilitates the understanding of scientific articles and that provides visualizations that complement the existing tools. However, this panel is only available in Portuguese. The tool is available at: http://simplificasus. com.

4 Conclusion

As the number of papers published in the literature continues to grow, machine learning techniques have become powerful tools to help us process and simplify knowledge. Here we investigated how sentiment analysis can be used to identify which technologies are presented as effective or not effective when used as a treatment option for a disease. In our case study, the pre-trained sentiment analysis model showed reasonable performance, indicating its potential as an early indication of the most promising therapies according to the literature. Thus, the use of machine learning and NLP can be helpful in summarizing information present in scientific articles and help guide such review efforts in the future. Finally, the manually curated database presented here can also serve as a basis for training more sophisticated models in the future, which may result in better tools for this task.

Acknowledgments. The authors would like to thank Mariana Parise for her valuable contributions during the discussions on the elaboration of this manuscript. The authors also thank the funding agencies: CAPES, CNPq, and FAPEMIG. We also thank the Campus Party, Brazilian Ministry of Health, Fiocruz, and DECIT teams.

Data Availability. Supplementary material, data, and scripts are available at https://github.com/ LBS-UFMG/SimplificaSUS.

References

1. Hu, B., Guo, H., Zhou, P., Shi, Z.-L.: Characteristics of SARS-CoV-2 and COVID-19. Nat. Rev. Microbiol. **19**(3), 141–154 (2021)
2. Pairo-Castineira, E., et al.: Genetic mechanisms of critical illness in COVID-19. Nature **591**(7848) (2021). Art. n° 7848. https://doi.org/10.1038/s41586-020-03065-y
3. Melms, J.C., et al.: A molecular single-cell lung atlas of lethal COVID-19. Nature **595**(7865) (2021). Art. n° 7865. https://doi.org/10.1038/s41586-021-03569-1

4. Pathak, G.A., et al.: A first update on mapping the human genetic architecture of COVID-19. Nature **608**(7921) (2022). Art. n° 7921. https://doi.org/10.1038/s41586-022-04826-7

5. Dos Santos, V.P., et al.: E-Volve: understanding the impact of mutations in SARS-CoV-2 variants spike protein on antibodies and ACE2 affinity through patterns of chemical interactions at protein interfaces. PeerJ **10**, e13099 (2022). https://doi.org/10.7717/peerj.13099

6. Harper, L., et al.: The impact of COVID-19 on research. J. Pediatr. Urol. **16**(5), 715–716 (2020). https://doi.org/10.1016/j.jpurol.2020.07.002

7. Glasziou, P.P., Sanders, S., Hoffmann, T.: Waste in covid-19 research. BMJ **369**, m1847 (2020). https://doi.org/10.1136/bmj.m1847

8. Fraser, N., et al.: Preprinting the COVID-19 pandemic. bioRxiv, p. 2020.05.22.111294, 5 de fevereiro de 2021. https://doi.org/10.1101/2020.05.22.111294

9. Painel de Evidências Científicas sobre Tratamento Farmacológico e Vacinas - COVID-19. https://infoms.saude.gov.br/extensions/evidencias_covid/evidencias_covid.html. acesso em 20 de abril de 2023

10. Li, X., et al.: Is hydroxychloroquine beneficial for COVID-19 patients? Cell Death Dis. **11**(7) (2020). Art. n° 7. https://doi.org/10.1038/s41419-020-2721-8

11. Schwartz, I.S., Boulware, D.R., Lee, T.C.: Hydroxychloroquine for COVID19: the curtains close on a comedy of errors. Lancet Reg. Health – Am. **11** (2022). https://doi.org/10.1016/j.lana.2022.100268

12. Avezum, Á., et al.: Hydroxychloroquine versus placebo in the treatment of non-hospitalised patients with COVID-19 (COPE – Coalition V): a double-blind, multicentre, randomised, controlled trial. Lancet Reg. Health – Am. **11** (2022). https://doi.org/10.1016/j.lana.2022.100243

13. Maisonnasse, P., et al.: Hydroxychloroquine use against SARS-CoV-2 infection in non-human primates. Nature **585**(7826) (2020). Art. n° 7826. https://doi.org/10.1038/s41586-020-2558-4

14. Hoffmann, M., et al.: Chloroquine does not inhibit infection of human lung cells with SARS-CoV-2. Nature **585**(7826) (2020). Art. n° 7826. https://doi.org/10.1038/s41586-020-2575-3

15. Dhibar, D.P., et al.: The 'myth of Hydroxychloroquine (HCQ) as post-exposure prophylaxis (PEP) for the prevention of COVID-19' is far from reality. Sci. Rep. **13**(1) (2023). Art. n° 1. https://doi.org/10.1038/s41598-022-26053-w

16. Ghazy, R.M., et al.: A systematic review and meta-analysis on chloroquine and hydroxychloroquine as monotherapy or combined with azithromycin in COVID-19 treatment. Sci. Rep. **10**(1) (2020). Art. n° 1. https://doi.org/10.1038/s41598-020-77748-x

17. Hutto, C., Gilbert, E.: Vader: a parsimonious rule-based model for sentiment analysis of social media text. apresentado em Proceedings of the International AAAI Conference on Web and Social Media, pp. 216–225 (2014)

18. Abubakar, A.R., et al.: Systematic review on the therapeutic options for COVID-19: clinical evidence of drug efficacy and implications. Infect. Drug Resist., 4673–4695 (2020)

19. Rahmani, H., et al.: Comparing outcomes of hospitalized patients with moderate and severe COVID-19 following treatment with hydroxychloroquine plus atazanavir/ritonavir. DARU J. Pharm. Sci. **28**(2), 625–634 (2020). https://doi.org/10.1007/s40199-020-00369-2

20. Gomez-Mayordomo, V., Montero-Escribano, P., Matías-Guiu, J.A., González-García, N., Porta-Etessam, J., Matías-Guiu, J.: Clinical exacerbation of SARS-CoV2 infection after fingolimod withdrawal. J. Med. Virol. **93**(1), 546–549 (2021). https://doi.org/10.1002/jmv.26279

21. Kim, Y.C., Dema, B., Reyes-Sandoval, A.: COVID-19 vaccines: breaking record times to first-in-human trials. NPJ Vaccines **5**(1), 34 (2020)

22. Karadeniz, H., Yamak, B.A., Özger, H.S., Sezenöz, B., Tufan, A., Emmi, G.: Anakinra for the treatment of COVID-19-associated pericarditis: a case report. Cardiovasc. Drugs Ther. **34**(6), 883–885 (2020). https://doi.org/10.1007/s10557-020-07044-3

23. Davis, M.R., McCreary, E.K., Pogue, J.M.: That escalated quickly: Remdesivir's place in therapy for COVID-19. Infect. Dis. Ther. **9**(3), 525–536 (2020). https://doi.org/10.1007/s40 121-020-00318-1

24. Roustit, M., Guilhaumou, R., Molimard, M., Drici, M.-D., Laporte, S., Montastruc, J.-L.: Chloroquine and hydroxychloroquine in the management of COVID-19: much kerfuffle but little evidence. Therapies **75**(4), 363–370 (2020). https://doi.org/10.1016/j.therap.2020. 05.010

25. La Rosée, F., et al.: The Janus kinase 1/2 inhibitor ruxolitinib in COVID-19 with severe systemic hyperinflammation. Leukemia **34**(7) (2020). . Art. n° 7. https://doi.org/10.1038/s41 375-020-0891-0

A Framework for Inference and Selection of Cell Signaling Pathway Dynamic Models

Marcelo Batista[1](✉) , Fabio Montoni[1,2] , Cristiano Campos[4] ,
Ronaldo Nogueira[1] , Hugo A. Armelin[2,3] , and Marcelo S. Reis[2,4](✉)

[1] Institute of Mathematics and Statistics, University of São Paulo, São Paulo, Brazil
`marcelobatista@ime.usp.br`
[2] CeTICS, Cell Cycle Lab, Butantan Institute, São Paulo, Brazil
[3] Institute of Chemistry, University of São Paulo, São Paulo, Brazil
[4] Recod Lab, Institute of Computing, University of Campinas, Campinas, Brazil
`msreis@ic.unicamp.br`

Abstract. Properly modeling the dynamics of cell signaling pathways requires several steps, such as selecting a subset of chemical reactions, mapping them into a mathematical model that deals with the communication of the pathway with the remainder of the cell (e.g., systems of universal differential equations - UDEs), inferring model parameters, and selecting the best model based on experimental data. However, this entire process can be extremely laborious and time-consuming for many researchers, as they often have to access different and complicated tools to achieve this goal. To address the challenges associated with this process in a more efficient way, we propose a framework that provides a streamlined approach tailored for universal differential equation UDE-based cell signaling pathway modeling. The open-source, free framework (github.com/Dynamic-Systems-Biology/BSB-2023-Framework) combines parameter inference algorithms, model selection techniques, and data importation from public repositories of biochemical reactions into a single tool. We provide an example of the usage of the proposed framework in a Julia Jupyter notebook. We expect that this streamlined approach will enable researchers to design improved cell signaling pathway models more easily, which may lead to new insights and discoveries in the study of biological mechanisms.

Keywords: Cell signaling pathways · Parameter inference · Biochemical reactions · Universal differential equation

1 Introduction

Model inference and selection are essential techniques in Systems Biology to obtain a mathematical model that adequately represents the studied biological phenomenon. In the context of cell signaling pathways, it is necessary to select a

M. S. Reis and R. C. de Melo-Minardi (Eds.): BSB 2023, LNBI 13954, pp. 82–93, 2023.
https://doi.org/10.1007/978-3-031-42715-2_8

subset of chemical reactions from a pathway for dynamic modeling, that is, the description of how the concentration of each chemical species in those reactions evolves over time. Dynamic modeling is often accomplished using ordinary differential equations (ODEs), and also requires model inference, that is, to adjust model parameters using experimental measurements. To model the dynamics of a cell signaling pathway, several steps are needed to be carried out:

- 1) To import a set of reactions from public repositories of systems biology databases such as BioModels [1] or REACTOME [10]. These reactions are usually accompanied by kinetic constants and initial concentrations and are described in either CSV (comma-separated values) or SBML (Systems Biology Markup Language) format [7].
- 2) To map the set of reactions into a mathematical model that describes the pathway dynamics, which is usually done using ordinary differential equations (ODEs). ODE equations describe how the concentrations of each chemical species in the signaling pathway change over time. This process usually involves defining state variables, kinetic equations for each reaction, and initial concentrations. There are several tools and software available to assist in this step, such as MATLAB [12], Python with the SciPy package [22], or specific software for modeling signaling pathways, such as COPASI [6].
- 3) To use inference algorithms to estimate the missing parameters of the mathematical model, such as kinetic constants and initial concentrations of some chemical species. There are several inference approaches available, including optimization-based approaches such as the least squares method [3], and simulation-based approaches such as the Monte Carlo algorithm [18].
- 4) Finally, if more than one set of reactions are considered for dynamic modeling, the model that best explains the dynamics of the cellular signaling pathway should be selected. This usually involves comparing the model with experimental data to verify its accuracy and validity. The criteria for selecting the model may vary depending on the modeling objective, but usually include the model's ability to accurately predict the behavior of the signaling pathway and the simplicity of the model.

However, the aforementioned classic signaling pathway modeling pipeline has an issue: selecting chemical species from a pathway for dynamic modeling implies in "disconnecting" these species from the remainder of the network, resulting in problems during the inference of model parameters. To mitigate this problem, estimating the missing communication due to the disconnection is necessary, but it is a computationally difficult problem and requires the estimation of extra parameters in the mathematical model. A recent way to infer that missing communication is the usage of ODE systems coupled with neural networks, which can be attained using mathematical frameworks such as the universal differential equations (UDEs) [17].

Currently, in the case of UDE mathematical models, the aforementioned sequence of steps typically requires researchers to manually execute a myriad of independent tools, which can be challenging, particularly for those who are not proficient in pipeline programming; such proccess can also be time-consuming and prone to errors. Therefore, there is a need for developing an streamlined approach tailored for UDE-based cell signaling pathway modeling that consolidates these steps into a single tool. Such approach should also consider the user-friendliness and allow the import of models from public reaction repositories. To address this need, we propose a new framework that combines data importation from public repositories of biochemical reactions, UDE-based signaling pathway model building, parameter inference and model selection, and simulation results analysis. This framework has the potential to lead to the generation of better cell signaling pathway models, which might assist in the studies of the underlying mechanisms of such pathways.

2   Related Frameworks

In this section, we present some existing applications with similar purposes, highlighting their limitations and demonstrating the need for a new framework to address these issues.

2.1   COPASI and Cell Designer

COPASI and Cell Designer [2] are both software tools used to simulate and analyze models of complex biological processes, including metabolic networks and cell signaling pathways. While COPASI enables the creation of mathematical models and the simulation of different experimental conditions to predict system behavior, it lacks intuitive model visualization and direct support for UDE systems. In contrast, Cell Designer provides a user-friendly graphical interface for creating models and visualizing molecule interactions, as can be seen in Fig. 1, but also lacks support for UDE systems and, akin to COPASI, may be time-consuming to use and require significant prior knowledge of the cell biology.

2.2   Garuda

Garuda is a software framework designed to analyze biological systems by integrating various types of biological data [20]. It offers a range of tools to construct and customize pathway models, and can work with multiple of these tools simultaneously, allowing researchers to build more comprehensive models. However, it should be noted that Garuda does not support UDE systems, and constructing biological systems models in Garuda may require significant prior knowledge of cell biology and other applications due to its complexity.

2.3   ABC-SysBio

ABC-SysBio is a command line-based framework for modeling biological systems that uses an Approximate Bayesian Computation (ABC)-based approach for model inference and selection [11]. It is capable of handling models of biological systems that involve multiple scales of time and size. However, ABC-SysBio can

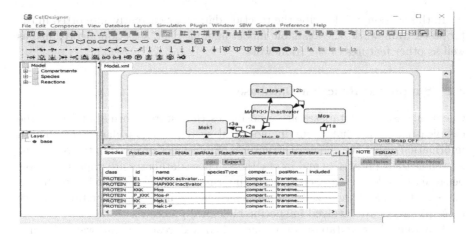

Fig. 1. Screenshot of Cell Designer software v.(4.4.2), showing an example of a biochemical model being built. Source: Cell Designer (www.celldesigner.org).

be complicated to configure and use, and may require specialized knowledge in programming for advanced modeling as the application does not have a graphical interface and requires the usage of command-line interfaces. Additionally, this framework was tailored for ODE-based dynamic models, not for UDE-based ones.

3   The Proposed Framework

We introduce a framework for model inference and selection of cell signaling pathways which allows for easy retrieving data from public reaction databases and usage of inference and model selection algorithms such as sequential heuristics [9], ABC methods [21], and others. The proposed framework was designed to ensure good usability, high computational performance, and also to overcome limitations of similar tools, such as providing support for UDE-based models. The source code of the framework is open and freely available at github.com/Dynamic-Systems-Biology/BSB-2023-Framework.

The proposed framework consists of a library containing various functions programmed in Julia and C++ programming languages. This was a strategic choice, as each of these languages offer unique advantages. C++ is known for its high performance and memory management capabilities, making it ideal for tasks that require intensive computation, such as numerical analysis and machine learning. Julia, on the other hand, is a language specifically designed for scientific computing, with a syntax that is both easy to read and write, and performance that rivals that of C++. Julia's built-in support for parallel computing and distributed memory architectures also makes it well-suited for applications that require scaling to larger datasets or distributed computing environments. Together, the use of these two languages enables the framework to be highly performant and versatile, with the ability to handle a range of tasks with ease.

In this section, we describe the reaction databases currently supported in the proposed framework, the available model inference and selection algorithms, as well as other features of the framework.

3.1   Supported Reaction Databases

Chemical reaction databases are repositories that store information about the chemical reactions that occur in biological systems. Those databases contain information about the chemical structures of molecules involved in the reactions, as well as their properties and interactions. Those databases are important tools because they allow researchers to access information about biochemical pathways in a structured and organized manner, which can be useful in understanding how these pathways are involved in specific biological processes.

Reactome is a database of human biological reaction pathways. It provides detailed information on biological processes such as cellular signaling, metabolism, and response to external stimuli, as well as showing the interactions between proteins, molecules, and other cellular components involved in these pathways. It is regularly updated with new information and used by researchers and scientists around the world to better understand the biological processes involved in disease and other areas of biology. In addition, Reactome has two main versions: a normal version, which is a relational database, and a graph version, which is a graph-based database. The normal version of Reactome is based on a traditional relational model, which is suitable for simple queries and data analysis with traditional data mining tools. On the other hand, the graph version of Reactome is based on a graph structure, which allows for the representation of complex relationships between the components of biological pathways. In this version, each biological component is represented as a node in the graph, while the interactions between these components are represented as edges between the nodes, as can be seen in Fig. 2. This graph representation is especially useful for more complex analyses, such as identifying protein subnetworks and visualizing complex interactions between proteins and other molecules. Both versions of Reactome are publicly available and can be accessed through the official Reactome website (reactome.org).

Recently, our research group has developed an extension to Reactome, named Anguix, that incorporates biochemical kinetic data into the models generated by the database [14,15]. These data, which consist of reaction rate constants, are imported from the SABIO-RK database, which has the relevant property of providing this information [23]. By incorporating reaction rate constants from SABIO-RK, the models generated by the Reactome database become more accurate and reliable; for this reason, we chose Anguix as the primary source of data in this project, though models imported from sources such as BioModels could also be used.

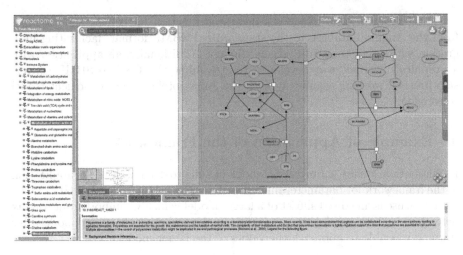

Fig. 2. Screenshot of Reactome pathway browser v.(3.7), showing an example pathway in the graph structure. Source: Reactome (reactome.org/PathwayBrowser/#/R-HSA-351202&PATH=R-HSA-1430728,R-HSA-71291).

3.2 Model Inference and Selection Algorithms

The framework enables parameter inference of models using one of the 13 different Julia optimization algorithms from the Optim.jl package [13]. Some of the available algorithms include gradient descent, Newton-Raphson, BFGS, and ADAM [19]. Each of these algorithms has its own advantages and disadvantages and is suitable for different types of problems. Gradient descent is a simple and widely used optimization algorithm, but it can lead to slow convergence or convergence to local minima. Newton-Raphson is an algorithm that takes into account the curvature of the objective function and can, therefore, converge more quickly than gradient descent, but it can be computationally expensive to compute the Hessian matrix in problems with many variables. BFGS is a more efficient version of gradient descent that approximates the Hessian matrix using information from the objective function. It is faster than Newton-Raphson and less susceptible to convergence to local minima. ADAM is a stochastic gradient-based optimization algorithm that is efficient on large datasets and exhibits good generalization capability.

For model selection, the proposed framework makes use of Bayesian methods, which can help to identify the most appropriate model for the data at hand. These methods utilize probability distributions from the Distributions.jl package to define priors, likelihoods, and posteriors, and the posterior distributions are then approximated using MCMC sampling methods, which are implemented in the Turing.jl package [5]. By comparing the posterior distributions of different models, we can select the one that best fits the data. Furthermore, we plan to include other selection algorithms in the future, such as cross-validation.

The framework also includes a dedicated function for converting CSV files from the Anguix database to the standardized SBML format, facilitating integration with other tools in Systems Biology. Additionally, an application with a graphical interface was developed to streamline the CSV to SBML conversion process, working seamlessly with the Julia notebook for convenient model conversion.

4   Example of Application of the Framework

In this section, we use a Jupyter notebook in Julia to demonstrate the application of the framework for parameter inference of a model using systems of UDEs. The model we use is an edited subset of a larger model, which we import from Anguix beforehand. Then, we plot the results and analyze whether the edited model after training was able to approach the values of the larger model.

4.1   Data Importation

First, we need to start the Neo4j application, which is a graph database management system [16]. Once the application is open, we can import data from the Anguix database by opening the executable, which can be downloaded from github.com/anthraxodus/Anguix-graphical (through this link we also find instructions for installing Neo4j and Reactome database), and selecting the organism we want to import the data from into Neo4j (e.g., *Danio rerio*). The sequence of screens of this process is showed in Fig. 3.

Fig. 3. Sequence of used interfaces for data acquisition from the Anguix database.

Once the data has been successfully imported into Neo4j, we can then perform queries using Cypher, a query language for querying graph databases created by Neo4j developers, to extract specific information from the database. In this case, our main model will consist of two reactions identified by the ID's 597 and 7489 from SabioRK database. This reactions involves the conversion of reactants into products through a series of chemical transformations. We can query for the model with these reactions in Neo4j and download it using the query shown in Eq. 1.

```
MATCH (n:SabioRkReaction)-[:kineticDataFor]-(k),
(n)-[:generalReactionFor]-(r), (k)-[:parameterInfo]-> (p)     (1)
RETURN n, k, r, p
```

Next, we will use the CSV to SBML conversion application to convert the model to SBML. This process can be seen in Fig. 4. After the conversion, we will place the resulting SBML file in the same folder as our Jupyter notebook `Pipeline.ipynb` named as `model.sbml`. Then, we will delete all reactions (and species involved in these reactions) except for "Reaction0" and "Reaction1", create a copy of the file, and repeat the same procedure, but this time we will leave only "Reaction1" and rename this copy as `cutmodel.sbml`. Finally, we will access the `Pipeline.ipynb` notebook.

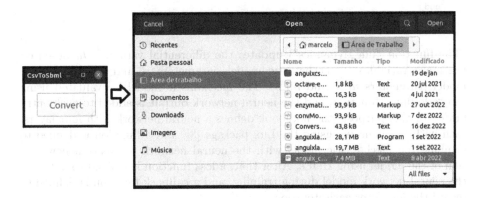

Fig. 4. Interfaces of the CSV to SBML conversion tool.

From this point, we just need to follow the Jupyter notebook block by block, making changes only to the names of files in the first blocks if necessary, and subsequently of the desired algorithms. Initially both models ("model" and "cutmodel") will be imported, and a simple simulation will be performed with its ODEs to generate the time series of the chemical species, as can be seen in the code snippet of Eq. 2.

```
# Generating the time series with a simulation
tspan = (0.0f0, 25.0f0); method = Rosenbrock23();
X, t = FWmodule.gen_timeseries(tspan, odesys, u0, model_param, ...
...method, abstol = 1e-12, reltol=1e-6, saveat = 0.1;
```
                                                                        (2)

This code block generates a time series for the given ODE system *odesys* by solving the system numerically using the "Rosenbrock23" method [4]. The simulation time span is defined by *tspan*, and initial conditions and model parameters are defined by $u0$ and *model_param*, respectively. The ODE problem is defined and solved internally by the "gen_timeseries" function using the method with specified tolerance and time step parameters. The resulting solution and the time points are stored in the variables X and t respectively.

Further in the notebook, there is a call to a function that performs the conversion of the ordinary differential equations (ODEs) from the model "cutmodel" to universal differential equations (UDEs), such function can be seen in the code snippet of Eq. 3.

```
function ude_dynamics!(du, u, p, nn_p, nn_st, t, ode_func, U)
    ode_func(du, u, p, t) # mechanistic model
    NN = U(u, nn_p, nn_st)[1] # add the neural network to the mechanistic model
    for i in 1:length(du)
        du[i] += NN[i]
    end
end
```

$$(3)$$

The function "ude_dynamics!" updates the differential variables du based on the state variables u, mechanistic model parameters p, and neural network model parameters nn_p and nn_st. The mechanistic model is evaluated using the "ode_func" function, and the neural network output is added to each entry in du. for this purpose, the notebook defines a neural network with four layers, each with ten neurons, using the Lux package [8], and a function is defined to combine the mechanistic model with the neural network to create a new version of the model using UDEs. After that, a loss function is defined to optimize the neural network model during training, and a callback function is defined to record the loss after each iteration.

Finally, The code block in Eq. 4 trains the neural network model using the ADAM optimization algorithm.

```
# Train with ADAM
_step = 1, losses = Float32[]; adtype = Optimization.AutoForwardDiff()
optf = Optimization.OptimizationFunction((x,p) -> loss(x), adtype)
optprob = Optimization.OptimizationProblem(optf, ComponentVectorFloat64(nn_p))
res1 = Optimization.solve(optprob,ADAM(0.1),maxiters=2000,callback=callback,progress=true)
loss_adam_end = size(losses)[1]
println(''Training loss after $(length(losses)) iterations: $(losses[end])'')
```

$$(4)$$

This procedure initializes a vector called *losses* to store the training loss after each iteration and defines an optimization function and problem based on the previously defined loss function. The ADAM solver is then used to optimize the neural network parameters, with the maximum number of iterations set by the *maxiters* parameter. The training loss is recorded after each iteration using the "callback" function. After training, the results are plotted, with Fig. 5 showing the loss function during training against the loss of the validation set and Fig. 6

showing the concentrations of selected species over time, simulated by the trained model (dashed lines) and the original model (solid lines). The predicted results closely match the actual ones.

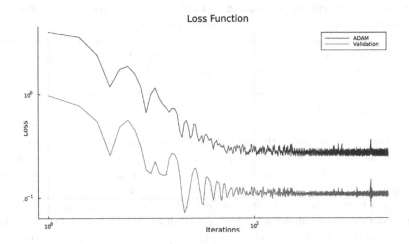

Fig. 5. Loss function evolution during training with ADAM.

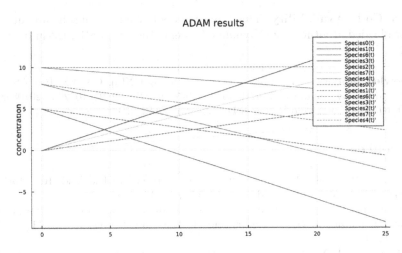

Fig. 6. Plot comparing the simulated results with the original model concentrations.

The files used in this notebook can be accessed at github.com/Dynamic-Systems-Biology/BSB-2023-Framework, thus allowing readers to reproduce the results and build upon the methods presented in this article.

For the purpose of comparison, when utilizing Copasi to obtain equivalent simulation results, a notable level of complexity becomes apparent. The proper implementation and configuration of the cellular signaling model in Copasi may require a profound understanding of the software and its modeling language. Additionally, constructing the model and defining its parameters can be a time-consuming process since it involves a manual procedure, taking several minutes depending on the complexity of the model. In contrast, our method automatically handles the accurate specification of differential equations and molecular interactions, ensuring precise and dependable results with greater ease.

5 Conclusions

In this paper, we reported the limitations of existing tools for inferring cellular signaling pathway models and proposed a new framework that overcomes these limitations. The new framework consists of a library of functions programmed in C++ and Julia, which allows: the conversion of data from the Anguix biochemical reaction database from CSV to SBML, parameter inference through optimization, model selection and results analyses. Some of those feature were demonstrated in this article through a Jupyter Julia notebook. Overall, we believe the further development of this framework might constitute an important contribution to the studies of biological systems, especially with regard to the inference of cellular signaling pathway models.

Source Code Availability. This framework is free and open software and can be downloaded at github.com/Dynamic-Systems-Biology/BSB-2023-Framework work.

Acknowledgements.. This work was supported by CNPq, CAPES, BECAS Santander and also by grants 13/07467-1, 19/21619-5, 19/24580-2, 20/10329-3, 20/08555-5 and 21/04355-4, São Paulo Research Foundation (FAPESP).

References

1. Chelliah, V., et al.: BioModels: ten-year anniversary. Nucleic Acids Res. **43**(D1), D542–D548 (2015). https://doi.org/10.1093/nar/gku1181
2. Funahashi, A., Morohashi, M., Kitano, H., Tanimura, N.: CellDesigner: a process diagram editor for gene-regulatory and biochemical networks. Biosilico **1**(5), 159–162 (2003)
3. Gutenkunst, R.N., Waterfall, J.J., Casey, F.P., Brown, K.S., Myers, C.R., Sethna, J.P.: Universally sloppy parameter sensitivities in systems biology models. PLoS Comput. Biol. **3**(10), e189 (2007). https://doi.org/10.1371/journal.pcbi.0030189
4. Hairer, E., Wanner, G.: Solving Ordinary Differential Equations. II: Stiff and Differential-Algebraic Problems, vol. 14. Springer, Heidelberg (1993). https://doi.org/10.1007/978-3-642-05221-7
5. Hong, C.W., Rackauckas, C.V.: Bayesian inference of dynamical systems using Turing.jl. J. Open Source Softw. **5**(47), 2067 (2020)

6. Hoops, S., et al.: COPASI-a complex pathway simulator. Bioinformatics **22**(24), 3067–3074 (2006)
7. Hucka, M., et al.: The Systems Biology Markup Language (SBML): language specification for level 3 version 2 core. J. Integr. Bioinform. **10**(2), 186 (2013). https://doi.org/10.1515/jib-2017-0081
8. Innes, M., Edelman, A., Fischer, K., Rackauckas, C., Saba, E.: Introducing Lux: a julia package for rapid development of custom deep learning models. J. Open Source Softw. **6**(57) (2021)
9. Ionita, M., Armeanu, A.: Sequential monte carlo methods for parameter inference in biological models. Int. J. Mol. Sci. **19**(12), 3811 (2018)
10. Joshi-Tope, G., et al.: The Reactome: a knowledge base of biologic pathways and processes. Nucleic Acids Res. **33**(Database Issue), D428–D432 (2005). https://doi.org/10.1093/nar/gki072
11. Liepe, J., et al.: ABC-SysBio-approximate Bayesian computation in Python with GPU support. Bioinformatics **26**(14), 1797–1799 (2010). https://doi.org/10.1093/bioinformatics/btq278
12. MathWorks: MATLAB (2021). Version R2021a
13. Mogensen, P., Larsen, A., Städler, N.: Optim.jl: a mathematical optimization package for Julia. J. Open Source Softw. **3**(24), 615 (2018). https://doi.org/10.21105/joss.00615
14. Montoni, F., et al.: Anguix: cell signaling modeling improvement through Sabio-RK association to Reactome. In: 2022 IEEE 18th International Conference on e-Science (e-Science), pp. 425–426 (2022). https://doi.org/10.1109/eScience55777.2022.00070
15. Montoni, F., et al.: Integration of Sabio-RK to the Reactome graph database for efficient gathering of cell signaling pathways, pp. 105–108 (2022). https://doi.org/10.5753/bresci.2022.222789. www.sol.sbc.org.br/index.php/bresci/article/view/20481
16. Neo4j Inc.: Neo4j (2007). www.neo4j.com/. Accessed 7 Apr 2023
17. Rackauckas, C., et al.: Universal differential equations for scientific machine learning. arXiv preprint arXiv:2012.09570 (2020)
18. Robert, C.P., Casella, G.: Monte Carlo Statistical Methods. 2nd edn. Springer, New York (2013). https://doi.org/10.1007/978-1-4757-4145-2
19. Ruder, S.: An overview of gradient descent optimization algorithms. arXiv preprint arXiv:1609.04747 (2016)
20. Smoot, M.E., Ono, K., Ideker, T.: Garuda and cyberinfrastructure: a recipe for interoperable and integrative analysis of complex data in the biological sciences (2010)
21. Toni, T., Welch, D., Strelkowa, N., Ipsen, A., Stumpf, M.P.H.: Approximate Bayesian computation scheme for parameter inference and model selection in dynamical systems. J. Roy. Soc. Interface **6**(31), 187–202 (2009). https://doi.org/10.1098/rsif.2008.0172
22. Virtanen, P., et al.: SciPy 1.0: fundamental algorithms for scientific computing in Python. Nat. Methods **17**, 261–272 (2020). https://doi.org/10.1038/s41592-019-0686-2
23. Wittig, U., et al.: SABIO-RK-database for biochemical reaction kinetics. Nucleic Acids Res. **40**(D1), D790–D796 (2012). https://doi.org/10.1093/nar/gkr1046

Intentional Semantics for Molecular Biology

Edward H. Haeusler[1]($\boxtimes$) , Bruno Cuconato[1] , Luiz A. Glatzl[1] ,
Maria L. Guateque[1] , Diogo M. Vieira[1] , Elvismary M. de Armas[1] ,
Fernanda Baião[2] , Marcos Catanho[3] , Antonio B. de Miranda[3] ,
and Sergio Lifschitz[1]

[1] Informatics Department, Pontifical Catholic University of Rio de Janeiro
(PUC-Rio), Rio de Janeiro, Brazil
{hermann,bclaro,lglatzl,mjaramillo,dvieira,earmas,sergio}@inf.puc-rio.br
[2] Industrial Department, Pontifical Catholic University of Rio de Janeiro (PUC-Rio),
Rio de Janeiro, Brazil
fbaiao@puc-rio.br
[3] Oswaldo Cruz Foundation (Fiocruz), Rio de Janeiro, Brazil
{mcatanho,amiranda,}@fiocruz.br

Abstract. This article presents an intentional semantics, using Object
Petri Nets (OPNs), to assign activity to each biological molecule and
complex, such as mRNA, tRNA, ribosomes, and protein synthesis. The
work differs from traditional uses of Petri Nets in Biology and Chemistry
for being a bottom-up and general semantics and not only a formalization
of some molecular biological phenomenon. Assigning activities to every
molecule and the difference between biological function and activity is
also a conceptual contribution of this work. To illustrate our semantics,
we set to tRNA, mRNA, ribosome, and the protein transcription molecular complex the respective activities expressed by OPNs.

1 Introduction

Any living organism must be able to store and preserve its genetic information,
transmit it to future generations, and express it as it carries out all of life's
processes. The main steps in handling genetic information are illustrated (Fig. 1)
by the Central Dogma of molecular biology [3]. As widely known, it represents
the flow of genetic information within a biological system.

According to the Central Dogma of molecular biology, the information
present in the DNA is transmitted to another molecule of DNA through a
process called replication (or duplication). This information is transferred to
RNA molecules through a process called transcription. The information in RNA
molecules is transmitted to protein (PTN) molecules through translation. We
also know that replication acts on the entire DNA of an organism, but not all
DNA is transcribed into RNA, and not all RNA is translated into proteins. We
want also to explore here, as seen in our graphical representation, that besides
DNA, RNA, and Protein interactions, there are also interactions with other
molecules at all levels.

M. S. Reis and R. C. de Melo-Minardi (Eds.): BSB 2023, LNBI 13954, pp. 94–105, 2023.
https://doi.org/10.1007/978-3-031-42715-2_9

All genetic information is abstracted using an alphabet of four nucleotides (A, G, T, and C) and is deciphered from sets of three nucleotides (codons); out of the 64 existing codons, 61 encode one of the 20 canonical amino acids, with the remaining three directing translation initiation and termination [4,13].

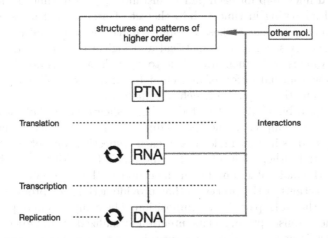

Fig. 1. A modified representation of the Central Dogma of Molecular Biology, with other interactions and molecules also represented.

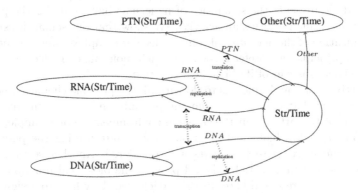

Fig. 2. The Set-theoretical Concepts and Mappings in Fig. 1

In Fig. 2, we provide a set-theoretical interpretation of the Central Dogma based on the same model. Besides RNA, DNA, Ribosomes, and other molecules explicitly mentioned in the model, molecules like polypeptides, ATPs, several polymers, and even simple molecules like Oxygen (O_2), metals, and so on, are also crucial in explaining the protein's activities. They are the *Others* set, i.e., other molecules, physically related domain objects, and even elements of concepts as time[1] Regarding a protein, activity is anything it can perform with any

---

[1] The time is not explicitly used in this paper. It appears here due to the completeness of the intended model.

proper molecule combination. The ellipses are sets of molecules; each solid line between sets represents a mathematical function that maps a molecule instance to the instantaneous time when it exists. For example, the *PTN* function represents the evolution in time of each molecule's individual. The dashed lines between solid lines map for each pair of instants $t_1 \preceq t_2$ in time the set of protein molecules existing in time t_1 into the set of protein molecules existing in time t_2. Some conditions on these mappings make them the mathematical concept of a sheaf. A sheaf is a geometrical and logical concept used to describe dynamical systems in mathematics. Due to space limits, this article focuses on a computational formalization of molecular biology, so we will not go too deep into the mathematical aspects of such a formalization.

Moreover, our big picture, depicted in Fig. 3, shows the level of the codes and data in the upper ellipse. In the lower ellipse, we show the denotation of each code (or process) as instant molecules in each respective concept set, i.e., RNA, PTN, DNA, and Other Molecules. The ribosome's transcription of the protein instant, i.e., the molecular complex in the center of the lower ellipse, is denoted by the compositions of the processes that provide activity or products by using the processes that take part in its composition. The composition of processes, as this ribosome's transcription, has its intention formalized by an Object Petri-Net (a concept explained in Sect. 3). The intuition is that it has a series of tRNA, the two parts of the ribosome, superior and inferior. The mRNA processes as inputs and delivers the protein transcription process as output.

This article presents the semantics of molecular biology under the principle that every molecule performs an activity. We show that our semantics supports the formalization of the activities of some molecular complexes and polypeptides. Under this principle, we prefer associating each molecule or molecular complex with its activity instead of its biological function.

The activity of a molecule has to do with all the possible chemical and physical interactions that it provides from the possible immediate combinations in which it can take part within the possible environments. For example, in contrast, a polypeptide's biological function involves interpreting the polypeptide activity whenever some real contexts are considered. Another semantic level, related to a forthcoming article, should provide the biological function concept to extend the one presented here. In this article, we show how to assign semantics for protein synthesis by using the semantics of its components. Our purpose is to convince the reader about the completeness of the model.

The paper is structured as follows. Section 2 gives an overview of computational semantics. Section 3 provides an informal introduction to the Object Petri Nets foundations. Section 4 defines the generalized model of object-nets, which

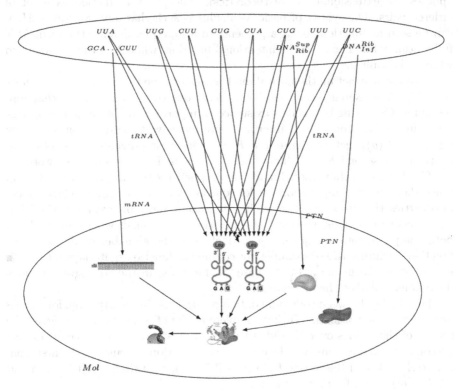

Fig. 3. The big picture of The intentional model, based on Fig. 1

allows a deep nesting structure to give semantics to molecular biology based on computational processes only. Section 5 describes related work and discusses them. The work closes with a conclusion.

2   Computational Semantics

A *model of computation* is a theoretical framework that describes how algorithms and computations can be performed.[2] It is a mathematical abstraction. For example, the Turing machine and finite state machines are models of computation that describe different classes of computations, the class of the former being a superset of the class pertaining to the latter.

Models of computation help analyze the capabilities and limitations of computers, programming languages, and algorithms. They also provide insight into some aspects of computation. For instance, the λ-calculus is an intentional model of functions. This means that functions are at the same level as arguments. This

---

[2] We must not confuse *model of computation* with *computational model*; the latter is a mathematical model of something that can be simulated or performed in a computer.

opposes the extensional, or set-theoretical, concept of a function as a set of ordered pairs. The syntax of basic λ-calculus is such that any two terms M, N can be seen as functions or as arguments to a function, making the term MN fully meaningful. The M is the intentional function, while N is the argument to which it is applied.

Even for abstract mathematical models, such as the theory of recursive functions [6], each recursive function has at least one code, or intention, that represents it. Observing this code's existence is important in obtaining the universal recursive function U that interprets any recursive function F on data d by $U(F, d) = F(d)$, such that, the first F is a code that provides an intention. In contrast, the second F is evaluated using the recursive functions framework.

The Turing machine model made this explicit for the first time in 1936 [14]. Since then, every computation model equivalent to the Turing machine model also carries the property of providing at least a code for every function. For the recursive functions, the code of a function is a number whose numeral can be decomposed into a sequence of characters in the alphabet of the recursive functions' definitions. For example, for coding the function that maps n to $n+2$, i.e., $s \circ s$, is the numeral 141 in base 20, for $s \circ s$, in the recursive functions definitions alphabet[3] has 20 symbols.

This duality between codes and data is natural and implicitly given for Turing machines. However, in any Turing-complete model of computation M, given the set Σ^* of the strings over Σ, and Λ the set of all strings that are codes for the M-computable functions, we always have two functions, $\mathcal{F}$ and $\mathcal{O}$, the first from Λ into $\Lambda \to \Lambda$, and the second from $\Lambda \to \Lambda$ into Λ, such that, $\mathcal{O}(\mathcal{F}(\Lambda)) \subset \Lambda$ and $\mathcal{F}(\mathcal{O}(\mathcal{F}(m)) = m$, for every $m \in \Lambda$.

In this article, we consider the following definition:

Definition 1. *A semantical-structure is any structure $\langle U, (\mathcal{F}_i)_{i \in \mathbb{N}}, \mathcal{O} \rangle$, such that, U is a set, for each $i \in \mathbb{N}$, $\mathcal{F}_i$ is a mapping from U into $U^i \to U$, and, $\mathcal{O}$ is a mapping from $\bigcup_{i \in \mathbb{N}} U^i$ into U, such that, (1) $\mathcal{O}(\mathcal{F}_i(U) \subset U$, $\mathcal{F}_i(\mathcal{O}(\mathcal{F}_i(m)) = m$, for every $i \in \mathbb{N}$, and (2) $\mathcal{F}_i(\mathcal{O}(\mathcal{F}_j(m)) = \mathcal{F}_{i-j}(m)$.*

In the following sections, we give a semantic structure for Molecular Biology, where U is the set of Molecular complexes from the point of view of the observable physicochemical process they induce. Due to the highly parallel and concurrent model of interaction of molecules with the environment and other molecules, we use Petri Nets, one of the most popular true concurrency[4] models.

3 Object Petri Nets Foundations

Petri Nets (PNs) can be seen as both a mathematical model and a graphical notation and are used as such in Computer Science and other areas. Carl Adam Petri

---

[3] The alphabet in question is $\{s, z, P, \circ, Rec_p, \langle, \rangle, \mu, \ldots\}$ with 20 letters.

[4] A true concurrency model can be taken as a framework to describe systems that allow many independent processes or threads to run simultaneously without interfering with each other model.

introduced them in 1962 [10] and have since been used for modeling, analyzing, and simulating dynamic systems exhibiting concurrency and synchronization. PNs are formally defined as a five-tuple (P, T, A, W, M_0) where P is a finite set of places; T is a finite set of transitions; $A \subseteq (P \times T) \cup (T \times P)$ is a set of arcs; $W : A \to \mathbb{N}$ is a weight function; $M_0 : P \to \mathbb{N}$ is a marking function.

The meaning of places and transitions in Petri Nets depends directly on the modeling approach. For example, in a dynamic system, a transition is regarded as an event, and the places are interpreted as a condition for an event to occur. The places contain tokens that travel through the net depending on the firing of a transition. These tokens simulate agents' dynamic and concurrent activities, instantiated from classes, and may be changed from place to place [2, 7].

The concept of Object-Based Petri Nets (OPNs) extends classical Petri Nets to represent and manipulate objects. In OPN, tokens are instances of classes defined as lists of attributes. Tokens, therefore, become a collection of constants, variables, net elements, and class elements that allow them to represent the object identifiers of sub-net instances, thereby allowing multiple levels of activity in the net and the dynamic allocation and deallocation of sub-nets. Object-Oriented Petri Nets (OOPNs) enhance OPNs with the notion of inheritance [9].

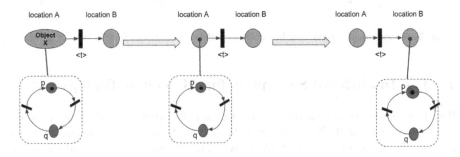

Fig. 4. Transport in an Object Petri Net. Use token "object X", then move object X

The most significant aspect of OPNs is the single unified class hierarchy. Providing a single unified class hierarchy in OPNs means that both token types and subnet types are classes, and these classes can be intermixed. A token can thus encapsulate a subnet. Therefore, OPNs support multiple activity levels on the net. The modeler is free to choose how various activities are to be composed, i.e., whether a particular object should be active, passive, or both depending on the focus of attention [8, 9].

For example, an OPN with two places (locations A and B) can model the movement of object X with a transition that is enabled by the token "object X", as shown in the first net in Fig. 4.

Moreover, as the object has a dynamical behavior – alternating states p and q – the token is again a marked net. A token net is also called an object net in distinction to the system net to which it belongs. The whole system is called an object net or short object system. The movement of the net token is shown as the firing of transition $\langle t \rangle$. Also, the token net can fire autonomously without being moved (see Fig. 5). Transport and autonomous firing can interleave; they are considered concurrent actions. This should be distinguished from a situation where these transition occurrences are synchronized, i.e., the object moves if and only if some object net transition occurs. Such an action may be triggered by the object, system, or both (the term interaction denotes this situation). A corresponding symbol labels interacting transitions [8,15], such as in Fig. 6.

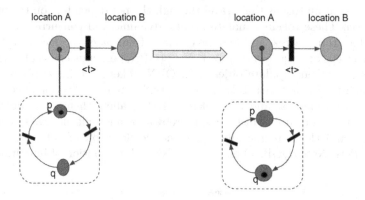

Fig. 5. Autonomous transition in an Object Petri Net model for protein synthesis

4   The Intentional Semantics for Molecular Biology

Based on the description of the Central Dogma of Molecular Biology, we propose an Object Petri Net (OPN) (see Fig. 7) for the modelling of the protein synthesis process. In this model, the mRNA, tRNA, amino acids, protein, and ribosome units are modelled as separate networks that work in parallel to the main network (places and transitions in grey in Fig. 7).

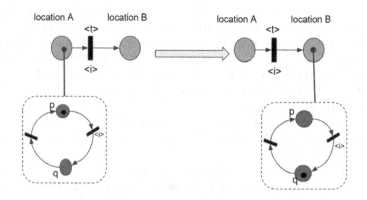

Fig. 6. Interaction in an Object Petri Net Model

The mRNA Petri (sub)Net is just a chain of places representing nucleotides as represented in Fig. 7. The transitions connecting them have labels synchronizing them with their matching tRNA, so the token will only transition with a token in a matching tRNA (i.e., containing the anticodon to the mRNA's codon).

A tRNA (sub)net is a 3-element chain of places representing nucleotides (see Fig. 7), followed by a single place representing the amino acid the tRNA carries. Its first three transitions have labels that synchronize with the transitions on the ribosome and the mRNA, ensuring that each nucleotide from the tRNA is processed by the ribosome as the corresponding nucleotide from the mRNA is processed.

The ribosome is the central piece of translation. The places labelled by mRNA, tRNA, and PTN have those nets as tokens (see Fig. 8).

A series of amino acid nets represent proteins. The token of each net is/represents the previous amino acid net as presented in Fig. 9. This process represents the step that joins amino acids together to form a protein. The amino acids in the figure are represented as the same amino acid. What is represented has to do only with joining the amino acids. The differences among the amino acids are related to the individual activity they have regarded in other contexts. Each has different activities; some help structure the body, while others regulate tasks. For each Petri net token in Fig. 9, the blank place must be filled with the activity the respective amino acid performs.

We recursively define a *Semantic-Structure* from the above constructions; see Definition 1. The intentions of basic molecules are the OPNs, mRNA, tRNA, ribosomes, and PTN. If $\mathbf{M} \in U$ is one of the basic molecules, $\pi_{\mathbf{M}}$ is the respective

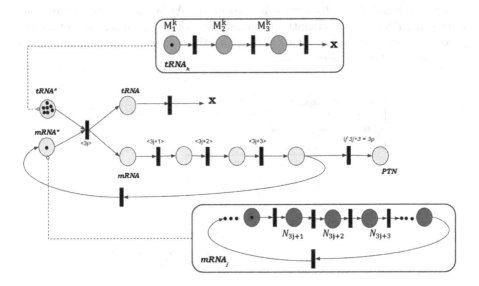

Fig. 7. Object Petri Net Model proposal for protein synthesis

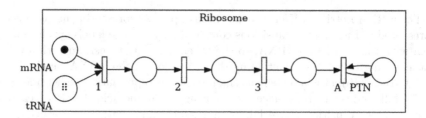

Fig. 8. Petri Net for Ribosome

OPN defined in this section and, $\mathcal{F}_{bpl(\pi_{\mathbf{M}})}(\mathbf{M}) = [\pi_{\mathbf{M}}]^5$. A marking m for any $\pi' \in [\pi_{\mathbf{M}}]$, corresponds to consuming the tokens in m and evolving to a process that corresponds to $\mathbf{M}$ reacting to the interactions in the level of bio-molecules. Let $\mathcal{O}([\pi]) = \mathcal{F}^{-1}_{pln(\pi)}(P([\pi]))$. Conditions (1) and (2) in Definition 1 hold, proving we have obtained a semantic structure by induction on the bio-molecule construction.

5 Related Works and Discussion

The usage of PNs to model processes is almost ubiquitous, ranging from theoretical to industrial applications. In [12], there is a comprehensive review on using PNs in Biology until 2013. Blätke et al. [1] also discusses PNs use in bio-applications, which include metabolic pathways, gene regulatory networks, and signalling pathways. Some applications are examples of how PNs can be used to analyze and simulate the behaviour of these systems.

Our work is different from what is cited above. As far as we know, there is no work using PNs to give semantics to molecular biology based on computational processes only. The above-mentioned applications concern the use of PNs to formalize specific processes.

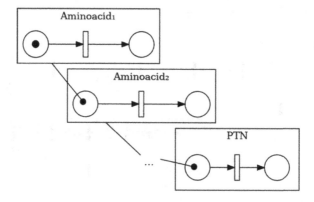

Fig. 9. Petri Net for PTN

---

[5] If π is an OPN, then $pl(\pi)$ is the number of output and input places in π, $[\pi]$ is the equivalence class of π under isomorphism on OPNs and P is the natural projection.

Recently, in [5], more sophisticated Petri Nets than OPNs are used, but they follow the approach we discuss in the sequence. As an illustrative example, Fig. 10 depicts a Petri-Net that describes the initial phase of the protein transcriptions. It starts with the TATAA code and consumes a series of *Transcription Factors*, TFIID, TFIIB, TFIIE, TFIIF, TFIIH in this order. Lastly, the kinase enzyme is implicitly consumed. Note that the PN asks, due to its coordination order, that, for example, TFIIF and the RNAPolyII need to be consumed simultaneously. The PN is opaque, since the tokens are not formalized, and the only information it provides is the high-level process it describes. Most PNs formalizations of Biological processes are of this kind. They choose some granularity of components to be taken as tokens and then describe the process top-down.

The semantics we provide follows in the opposite direction. The granularity is the lowest possible, starting with nucleotides and even simpler molecules. The focus is on the description of a broader scenario. For example, adding an amino acid to a growing peptide chain requires four ATP molecules, two for amino acid activation and two ATPs more for peptide bond formation and ribosome translation, see [11].

This is depicted in Fig. 11, the redesigned OPN that adds amino acids pushed by ATPs in the growing protein chain. We are not considering the additional costs of other ATPs for error correction and the synthesis of sequences that are removed during protein maturation since this has to do with protein maturation, which occurs after what we consider here. We also do not consider the two ATPs used in the translation phase for similar reasons. It is natural to include an energy promoter for bio-systems, for example, the *ATP*, as an object token in the intention of a process. By OPN-*transport*, the *ATP* goes to a position that it can be used in association with the pairs of nucleotide processes that it helps to implement, as illustrated in Fig. 11. This can culminate with allowing of describing self-sustainable consumption and generation of resources and energy usage.

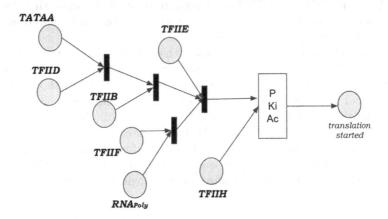

Fig. 10. High level Petri-Net for Transcription Starting Process

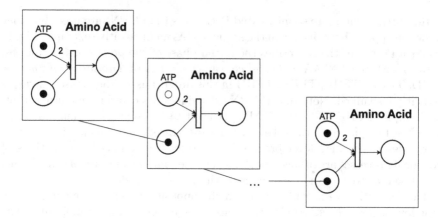

Fig. 11. Petri Net for PTN Growth with energy. 2 means two ATPs

Summing up, our model provides semantics based on process and is thus intentional (in the logical sense of the word). It is not a sole formalization that aims to predict some specific phenomenon. The semantics of transcription obtains by an OPN that has as tokens processes defined by OPNs as in Sect. 4.

6   Conclusion

This article presents the semantics for molecular biology under the principle that every molecule performs an activity. The semantics formalizes the activities of each molecular complex and polypeptides intentionally using OPNs. The OPN describes for each molecule all the possible chemical and physical interactions that it provides from the possible immediate combinations in which it can take part within any possible environment. We provided basic biological examples, RNA, Ribosome, tRNA, mRNA, and protein synthesis to show that we can, compositionally, assign semantics to more sophisticated molecular complexes from these basic intentional processes.

A possible further work is to associate the activity of each molecular complex with the biological function by interpreting this activity when some real contexts are considered. We expect that other semantic levels should provide this biological function concept to extend the one presented here.

This article has laid the foundational work that will allow us to formalize a wide class of biological processes. We have started with a formalization for protein synthesis. We plan to work it out in detail for specific proteins, starting from a simpler peptide like Insulin and moving on to more complex proteins afterward. We then intend to formalize other biological processes.

References

1. Blätke, M.A., Rohr, C., Heiner, M., Marwan, W.: A petri-net-based framework for biomodel engineering. In: Benner, P., Findeisen, R., Flockerzi, D., Reichl, U., Sundmacher, K. (eds.) Large-Scale Networks in Engineering and Life Sciences. MSSET, pp. 317–366. Springer, Cham (2014). https://doi.org/10.1007/978-3-319-08437-4_6
2. Chaudhuri, P.P., Ghosh, S., Dutta, A., Choudhury, S.P.: A New Kind of Computational Biology: Cell Automata Based Models for Genomics and Proteomics (2018)
3. Crick, F.: Central dogma of molecular biology. Nature **227**(5258), 561–563 (1970)
4. Diercks, C.S., Dik, D.A., Schultz, P.G.: Adding new chemistries to the central dogma of molecular biology. Chem **7**(11), 2883–2895 (2021)
5. Herajy, M., Liu, F., Rohr, C., Heiner, M.: Coloured hybrid petri nets: an adaptable modelling approach for multi-scale biological networks. Comput. Biol. Chem. **76**, 87–100 (2018)
6. Kleene, S.C.: λ-definability and recursiveness. Duke Math. J. **2**(2), 340–353 (1936)
7. Koch, I., Reisig, W., Schreiber, F.: Modeling in Systems Biology: The Petri Net Approach, vol. 16. Springer, London (2010). https://doi.org/10.1007/978-1-84996-474-6
8. Köhler, M., Rölke, H.: Properties of object petri nets. In: Cortadella, J., Reisig, W. (eds.) ICATPN 2004. LNCS, vol. 3099, pp. 278–297. Springer, Heidelberg (2004). https://doi.org/10.1007/978-3-540-27793-4_16
9. Lakos, C.A.: The object orientation of object Petri nets. Department of Computer Science, University of Tasmania (1995)
10. Petri, C.A.: Kommunikation mit automaten. Ph.D. thesis, Math. Uni. Bonn (1962)
11. Piques, M., et al.: Ribosome and transcript copy numbers, polysome occupancy and enzyme dynamics in Arabidopsis. Mol. Syst. Biol. **5**(1), 314 (2009)
12. Singh, N., Singh, A., Singh, S., Kumar, V.: Modeling and analysis of biological systems using petri nets: a review. J. Theor. Biol. **335**, 94–105 (2013)
13. Tan, C.L., Anderson, E.: The new central dogma of molecular biology. Resonance **14**(3), 1–32 (2020)
14. Turing, A.M.: On computable numbers, with an application to the Entscheidungsproblem. Proc. London Math. Soc. **s2-42**(1), 230–265 (1937)
15. Valk, R.: Object petri nets. In: Desel, J., Reisig, W., Rozenberg, G. (eds.) ACPN 2003. LNCS, vol. 3098, pp. 819–848. Springer, Heidelberg (2004). https://doi.org/10.1007/978-3-540-27755-2_23

transcAnalysis: A Snakemake Pipeline for Differential Expression and Post-transcriptional Modification Analysis

Pedro H. A. Barros[1,2(✉)] [ID], Waldeyr M. C. Silva[1] [ID],
and Marcelo M. Brigido[1] [ID]

[1] Department of Cellular Biology, Institute of Biology,
University of Brasilia, Brasília 70910-900, Brazil
[2] Graduate Program in Molecular Biology, University of Brasilia, Brasília, Brazil
`pedroa_barros1@hotmail.com`

Abstract. The transcAnalysis pipeline is a comprehensive tool that allows the analysis of transcriptome data. The pipeline allows for analysis of differential expression, alternative splicing, lncRNA and RNA editing analysis, with a specific focus on A-to-I editing mediated by the ADAR protein. This type of RNA editing is widespread and can significantly affect gene regulation and function. The results from these analyses are integrated, and the events are associated with each gene. The pipeline also integrates results that can help correlate gene expression and post-transcriptional events. This allows for a comprehensive understanding of the functional impact and provides insight into the biological processes and pathways associated with these events. One of the significant advantages of the transcAnalysis pipeline is its ability to perform all these analyses with a single command using the Snakemake package. This feature simplifies the analysis process and makes it accessible to researchers with limited bioinformatics expertise. Its user-friendly ability to perform multiple analyses with a single command make it an ideal choice for researchers looking to analyze transcriptome data.

Keywords: Pipeline · Differential Gene Expression · Alternative Splicing · RNA Editing

1 Introduction

The advent of next-generation sequencing (NGS) technologies has brought about a wealth of data that can be used to gain insights into biological systems. RNA sequencing (RNA-seq) has become the cornerstone for studying gene expression, with a primary focus on differential expression analysis [1]. However, this approach often overlooks other important information related to post transcriptional modifications and the expression of long non-coding RNAs (lncRNAs), which can also be included in the analysis.

M. S. Reis and R. C. de Melo-Minardi (Eds.): BSB 2023, LNBI 13954, pp. 106–111, 2023.
https://doi.org/10.1007/978-3-031-42715-2_10

One of the critical post-transcriptional modifications is alternative splicing (AS), which involves the removal of introns and non-canonical joining of exons from pre-mRNA, leading to the production of different proteins and transcriptional control. There are five primary forms of AS: exon skipping (ES), alternative 5' splice site (A5SS), alternative 3' splice site (A3SS), mutually exclusive exons (MXE), and intron retention (IR) [2]. These processes play a critical role in regulating gene expression and contribute to the diversity of the proteome.

RNA editing (RED) is mainly carried out by the protein ADAR in mammals, leading to the editing of adenine to inosine (A-to-I), which modifies the secondary structure of transcripts and alters the binding of RNA-binding proteins and miRNAs. RED can also lead to the generation of transcript isoforms through AS. These events are related to specific conditions and can aid in understanding biological phenomena [3]. By considering AS and RED in RNA-seq analyses, researchers can gain a more comprehensive understanding of the transcriptional and post-transcriptional mechanisms underlying gene expression.

2   Material and Methods

2.1   Workflow

The transcAnalysis pipeline performs mRNA, lncRNA, AS, and RE expression analysis from a BAM (Binary Alignment Map) file created after the alignment of RNA-seq reads from fastq files. Additionally, the pipeline requires a metadata file that is filled out by the user with their desired preferences to be executed, including the path to each sample. The pipeline facilitates the analysis by allowing the execution of each step with only one command line, which is possible due to the use of the snakemake, a workflow management system that provides integration with Conda and Docker, two popular tools in the bioinformatics community, to enhance reproducibility and portability of analysis pipelines [4]. The pipeline outputs the data integration related to the analyzed event (Fig. 1) and is available at: github.com/PHAB1/transcAnalysis.

2.2   Transciptome Analysis Pipeline

From the output obtained by the pipeline, the integration of the data related to each gene simultaneously to differential expression, RED (A-to-I), and AS is performed, considering each AS category separately (Table 1).

Differential Expression. Differential expression analysis was performed using the R package DESeq2[1] (v1.38.2), developed by Bioconductor project. The program specializes in normalization, visualization, and differential gene expression analysis, and uses empirical Bayes techniques to estimate the log of fold change, the dispersion, and related estimates such as the p-value and the False Discovery Rate (FDR) [5]. The program uses the gene expression array and a table containing the experimental design and performs differential expression analysis from two or more different conditions.

---

[1] https://bioconductor.org/packages/release/bioc/html/DESeq2.html.

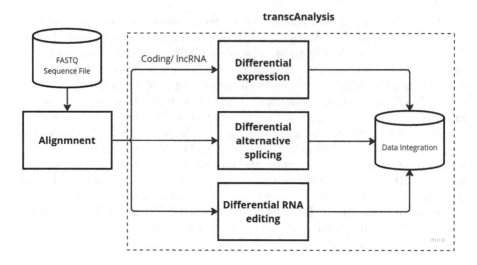

Fig. 1. Workflow and steps in the transcAnalysis pipeline

Table 1. Integration of transcriptome data. Only Fold change (FC) of which had $FDR < 0.05$ are shown. Similarly, the event counts of Exon Skkiping (ES), Intron Retention (IR) and RNA editing shown are significant with $FDR < 0.05$.

Gene	FC	SE	RI	...	RED
CTSS	0	4	0		42
APOBEC3C	0	0	0		13
METTL7A	0	0	0		16
⋮	⋮	⋮	⋮		⋮
IFITM2	2.35	0	0		3
RBM39	0	8	1		2

Alternative Splicing. For detection of differential AS, the rMATS[2] turbo program (v4.1.2) was used. The program is based on using mapped reads in regions indicating different isoforms to detect and estimate the proportion of different types of splicing between different conditions. Here, the five possible splicing types are identified, these being ES, IR, MXE, A3SS, and A5SS [6]. As an example, in ES, the reads mapped in the regions used as AS markers are the junction reads (S), positioned between the upstream exon and the downstream exon with single exon skipping, forming the isoform, while the reads related to the canonical form of the gene, with the inclusion of the exon, are the inclusion reads (I). Inclusion levels (ψ) are defined as the percentage of transcript inclusion [6].

$$\psi = (I/LI)/(I/LI + S/LS),$$

---

[2] https://github.com/Xinglab/rmats-turbo.

where:

ψ = Inclusion Level (IncLevel),

I = number of reads mapped to the inclusion isoform,

S = number of reads mapped to the ES isoform,

LI = effective length of the inclusion isoform exon,

LS = effective length of the ES isoform.

For filtering of the splicing events, after identification with rMATS, $FDR <$ 0.05 and $|\psi_{it} - \psi_{ic}| > 0.1$ were used, where ψ_{ic} and ψ_{it} indicates the relative mean ψ in each event between the (t) treatment and (c) control.

RNA Editing. For the detection of RED, the program SPRINT[3] (v0.1.8) was used. The SPRINT program preprocesses the reads by removing annotated single nucleotide genetic variants (SNPs), aiming to remove false positives from genetic variants, and subsequently identifies ER candidates [7]. Only candidates containing nucleotides with high quality $q > 20$ and mapped reads in regions with indistinguishable repeats are retained, removing those with mapping or sequencing errors.

LncRNA. Annotation of lncRNAs is performed and differential expression analysis is done with the DESeq2 program. For pathway enrichment analysis, the R package LncRNAs2Pathways[4] (v.1.1), developed to associate pathways related to lncRNAs, is used. The package uses a network library, which relates lncRNAs to the expression of mRNAs [8]. The interaction network was created from the analysis of 28 different studies and takes into account protein-protein interactions annotated in different databases such as REACTOME [9] and HPRD [10], and is able to relate differentially expressed lncRNAs to "KEGG" [11] or "Reactome" [9] pathways.

2.3 Experimental Transcriptome Data

The transcriptome samples in fastq format were obtained from the Sequence Read Archive (SRA) database. The experimental files used were all paired-end files. We used monocyte samples from patients with severe-stage COVID-19 (PRJNA699856). 6 treated and 6 controls were used. The STAR[5] program (v2.7) was used to align each sample to the reference genome GRCh38 (hg20) [13].

3   Results and Discussion

The pipeline performs statistical analysis between two distinct groups, including differential analysis of gene expression, lncRNA, AS, and RED. The pipeline is

---

[3] https://github.com/jumphone/SPRINT.

[4] https://cran.r-project.org/web/packages/LncPath/.

[5] https://github.com/alexdobin/STAR.

designed to integrate these events, allowing a complementary analysis to conventional expression analysis, as shown in Fig. 2, where the intersection of enriched terms between differentially expressed genes, differential alternative splicing, and differentially edited RNAs in monocytes from patients with severe COVID-19 versus the healthy patient is shown.

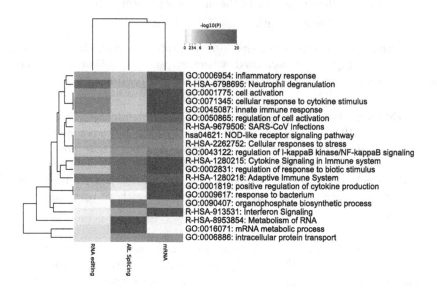

Fig. 2. Metascape functional analysis Heatmap of the transcriptome. Intersection of the most relevant terms in mRNA-related enriched genes, alternative splicing or RNA editing events.

3.1   Conclusion

There is a large amount of information in transcriptome data that is not normally used, programs that identify and analyze post-transcriptional modifications have no trivial use and require computer skills. The transcAnalysis pipeline was created with the intention of allowing the acquisition of data related to both gene expression and post-transcriptional modifications for the utilization of the data and integration, allowing association between the events. In addition, the Snakemake pipeline manager was used to create a user-friendly approach.

References

1. Hardwick, S., Deveson, I., Mercer, T.: Reference standards for next-generation sequencing. Nat. Rev. Genet. **18**(8), 473–484 (2017)
2. Marasco, L.E., Kornblihtt, A.R.: The physiology of alternative splicing. Nat. Rev. Mol. Cell Biol. **24**(4), 242–254 (2023)

3. Song, B., Shiromoto, Y., Minakuchi, M., Nishikura, K.: The role of RNA editing enzyme ADAR1 in human disease. WIRES **13**(1), e1665 (2023)
4. Mölder, F., et al.: Sustainable data analysis with Snakemake. F1000Res **18**, 10–33 (2021)
5. Love, M.I., Huber, W., Anders, S.: Moderated estimation of fold change and dispersion for RNA-Seq data with DESeq2. Genome Biol. **15**(12), 550 (2014)
6. Shen, S., et al.: rMATS: robust and flexible detection of differential alternative splicing from replicate RNA-Seq data. Proc. Natl. Acad. Sci. U.S.A. **111**(51), E5593–E5601 (2014)
7. Zhang, F., Lu, Y., Yan, S., Xing, Q., Tian, W.: SPRINT: an SNP-free toolkit for identifying RNA editing sites. Bioinformatics **33**(22), 3538–3548 (2017)
8. Han, J., et al.: LncRNAs2Pathways: identifying the pathways influenced by a set of lncRNAs of interest based on a global network propagation method. Sci. Rep. **7**, 46566 (2017)
9. Gillespie, M., et al.: The Reactome pathway knowledgebase 2022. Nucleic Acids Res. **50**(D1), D687–D692 (2022)
10. Mishra, G.R., et al.: Human protein reference database-2006 update. Nucleic Acids Res. (34), D411–D414 (2006)
11. Kanehisa, M., Furumichi, M., Tanabe, M., Sato, Y., Morishima, K.: KEGG: new perspectives on genomes, pathways, diseases and drugs. Nucleic Acids Res. **45**(D1), D353–D361 (2017)
12. Nishimura, D.: BioCarta. Biotech Softw. Internet Rep. **2**(3), 117–120 (2001)
13. Dobin, A., et al.: STAR: ultrafast universal RNA-Seq aligner. Bioinformatics **29**(1), 15–21 (2013)

Peptide-Protein Interface Classification Using Convolutional Neural Networks

Lucas Moraes dos Santos[✉] [iD], Diego Mariano[iD], Luana Luiza Bastos[iD], Alessandra Gomes Cioletti[iD], and Raquel Cardoso de Melo Minardi[iD]

Laboratory of Bioinformatics and Systems Institute of Exact Sciences, Department of Computer Science, Federal University of Minas Gerais, Belo Horizonte, Brazil
moraes.lsantos@gmail.com,
lucasmds@ufmg.br{lucas.santos,raquecm}@dcc.ufmg.br

Abstract. Peptides are short chains of amino acid residues linked through peptide bonds, whose potential to act as protein inhibitors has contributed to the advancement of rational drug design. Indeed, understanding the interactions between proteins and peptides is potentially helpful for several biotechnological applications. However, it is not a trivial task since peptides can adopt different conformations when interacting with proteins. In this paper, we develop a classification model for protein-peptide interfaces using a convolutional neural network and distance maps. To evaluate our proposal, we performed two case studies classifying protein-peptide interfaces based on peptide sequences and receptor classes. Additionally, we compared the distance map approach with a graph-based structural signatures approach. We aim to find out if a convolutional neural network could classify peptides just from the patterns of distances in these maps. In conclusion, graph-based methods were slightly superior in almost all comparisons performed. However, distance map-based signature methods achieved better results for some classes, such as classifying hormones, membranes, and viral proteins. These results shed light on the potential use of distance maps for classifying protein-peptide interfaces. Nevertheless, more experiments may be needed to explore this use.

Keywords: Convolutional neural networks · distance maps · protein-peptide interactions

1 Introduction

Peptides are short-chain molecules consisting of two to fifty amino acid residues linked through peptide bonds. They have several essential functions in human physiology, such as acting as hormones, neurotransmitters, growth factors, ion channel ligands, or anti-infective agents [23]. Moreover, recent research suggests that peptides play a vital role in protein-protein interactions, constituting a significant percentage of such interactions within cells [2].

Compared to proteins, peptides have more chemical versatility because they can be more easily modified. Additionally, peptides exhibit low resistance and

M. S. Reis and R. C. de Melo-Minardi (Eds.): BSB 2023, LNBI 13954, pp. 112–122, 2023.
https://doi.org/10.1007/978-3-031-42715-2_11

limited non-target activity, making them suitable for therapeutic agents [14, 28]. As a result, peptide drug development has become one of the hottest topics in pharmaceutical research.

Designing new peptides and peptide-based compounds for drug development and biotechnological applications requires understanding the structure and recognition of protein-peptide complexes. With the aid of databases containing protein-peptide complexes, researchers can analyze and gain insights into the mechanisms of protein-peptide recognition, paving the way for future discoveries [5, 16]. However, these studies depend on public structure databases, such as PDB (Protein Data Bank) and more specialized databases, as Propedia [19].

Propedia is a database of peptide-protein interactions designed to provide a comprehensive and current dataset of complex protein-peptide experiments [18]. In a recent study [19], graph-based structural signatures [17] have been used to extract characteristics of protein-peptide complexes collected from Propedia. Then, several machine-learning approaches were used to classify protein-peptide complexes [19]. The results demonstrated the potential use of graph-based signatures for protein-peptide classification. However, other approaches could be used to construct new signature types, such as distance maps.

Mathematical approaches applied to understanding the properties of proteins have provided insights relevant to structural bioinformatics [13]. For example, information about the structure of biomolecules is encoded in the internal distances, represented by square matrices known as distance matrices. These representations contain the pairwise distances between residues in a protein and are used to infer protein-protein interactions [13].

A distance matrix can be defined as $\mathbf{d} = (d_{ij})$, where d_{ij} is the Euclidean distance between the ith and the jth residue. Generally, the coordinates of the atoms of C_α (carbon-α) and/or C_β of the residuals are input to the method [13].

Recent studies have shown that predicting the structures of proteins can be done using two-dimensional images known as distance maps (DMs), representing the inter-residue distance matrices of proteins. These maps are increasingly used to compare biomolecular structures and analyze functional differences between proteins [11]. By comparing DMs of homologous structures, researchers can identify similarities and differences in their patterns of structural flexibility [11]. In addition, DMs have the added advantage of being low-dimensional, invariant to rotation and translation of structures, making parameter calculation and efficient learning [6], which is desirable for artificial intelligence applications, such as convolutional neural networks.

Convolutional neural networks (CNNs) [15] are a class of deep neural networks, of the type *feed-forward*, specialized in processing data that have a topology of *grid* (*e.g.*, image) [9]. The architecture of a CNN is analogous to the connectivity pattern of neurons in the human brain being inspired by the organization of the visual cortex, where neurons in different layers detect increasingly complex features of visual stimuli. As an allusion to their name, these neural networks use a mathematical operation called *convolution* in feature learning, as opposed to matrix multiplication common in multilayer perceptrons (MLP) [9].

This class of neural networks has shown great potential in applications involving pattern recognition in images [9], being used recently in conformational analysis, structure prediction, protein classification, etc. [21]. The basic structure of CNNs consists basically of two parts: feature learning (convolution and pooling layers) and classification (fully-connected layers) [21].

In this study, we model the interface region of the protein-peptide complex through a two-dimensional representation of the interatomic distance matrix, known as a distance map. We aim to find out if a CNN could classify peptides from the patterns of distances in these maps. The importance of this approach in the context of Bioinformatics/Biotechnology stems from its contribution to the advancement of computational modeling techniques for analyzing biological data. By improving modeling capabilities, we enable more effective machine learning applications across multiple scenarios, enabling researchers to gain deeper insights, make more accurate predictions, and accelerate advances in understanding and harnessing biological systems.

2   Material and Methods

2.1   Data Collection

The protein-peptide complexes used in this work come from the Propedia web database (http://bioinfo.dcc.ufmg.br/propedia2). We performed two case studies. First, we analyzed 1,111 peptides from five clusters grouped by sequence similarities (clusters S0, S1, S112, S151, and S162). Additionally, we collected and analyzed 6,238 peptides from six Propedia datasets: AntimicrobialDB, ViralDB, EnzymeDB, MembraneDB, HormoneDB, and PlantDB. Lastly, we compared our results to the neural network analysis of graph-based signatures shown in [19] (signature method: aCSM-ALL with 0.2Å of step and distance max of 20Å [25]; parameters used in Orange Data Mining [7]: neurons in hidden layers = "300", solver = "Adam", activation = "ReLu", maximal number of iterations = "200", regularization alpha = "0.001", and replicable training).

2.2   Generation of Distance Maps

We focus on atoms within the interface region to generate distance maps for protein-peptide complexes. We select the residues C_α (alpha carbon) from each .pdb file and extract their corresponding coordinates (x, y, z) from the protein and peptide structures. Using these coordinates, we calculate the Euclidean distance between atoms within the interface region to create a distance matrix between residues. In this matrix, peptide atoms correspond to the ordinate axis, while protein atoms correspond to the abscissa axis [20]. Finally, we transform the distance matrix into a two-dimensional image (*.png* format) using Python's Matplotlib library. The algorithms for developing the process described above and obtaining the distance maps were developed in Python (version 3.7.9)[1].

---

[1] https://github.com/LBS-UFMG/cnn-distance-maps.

Distance maps offer a powerful means for inferring three-dimensional structure using paired distances. By analyzing attention maps, for example, we can effectively identify significant patterns between residual pairs. Furthermore, as mentioned earlier, the main advantage of distance maps lies in their inherent invariance to rotations and translations of the protein structure. The computational complexity of this method is denoted as $O(n \times m)$, where n represents the number of residues in the protein and m denotes the number of residues in the peptide. However, when distance maps are utilized as input for CNNs, additional pre-processing steps, such as resizing, are often required, particularly if the distance matrix is not square.

2.3   Pre-processing: Data Augmentation and Resizing

To prepare the DMs for input into our neural network, we applied preprocessing techniques that consisted of three steps: resizing, data augmentation, and rescaling. Since the protein-peptide complex can contain molecules of varying size, we need to resize the DMs to 64×64 pixel dimensions to fit the square input structure required by the CNN architecture. Following this, we applied data augmentation to the DM set using a series of techniques such as brightness adjustments, sharpening filters, and horizontal/vertical shifts. Previous studies have shown that data augmentation techniques can significantly improve classification models, particularly for imbalanced datasets [30]. Lastly, we rescaled each DM so that the pixel values were converted to a range between 0 and 1 since neural networks tend to perform better with values in this range [9].

2.4   Model Architecture

Our model is based on representation learning [3], a technique that allows the system to automatically learn important features from a large amount of data, allowing it to learn a representation specific to the task. We use a popular representational learning technique called Deep learning that involves using deep neural networks that optimize weight parameters, by combining simple and complex features to construct hierarchical representations of input data [21]. We employ a type of deep neural network called *Convolutional Neural Networks* (CNN).

We developed a sequential and uniform architecture [4] comprising a linear stack of 2D convolutional layers. The first two layers have 32 filters each, while the last two have 64 filters each. We define a 3×3 convolution *kernel* with a stride of 1. After the convolutional layers, we include a pooling layer with max-pooling using a 2×2 pool array and a stride of 2. To enhance the nonlinear properties of the feature maps generated, we apply the Rectified Linear Unit (ReLU) activation function [9], which is followed by a Batch Normalization layer to zero center the activations [10].

For our model, we selected a batch size of 32, which determines the number of samples processed by the network in one pass. Typically, larger batch sizes demand more memory, so it's common to use values like 32 or 64 [22]. The input data was structured as *tensors*, defined by the input shape, which includes the image dimensions (height and width) and the number of color channels (RGB is equivalent to 3). Additionally, the batch size was specified [4].

We generate the input layer for the fully connected layers (FC layer) by vectorizing the feature maps and concatenating them into a flattened array. For multiclass classification, the FC layer acts as a classifier with 512 nodes, and we employ the softmax activation function to process its output [27]. Additionally, to improve generalization and prevent overfitting, we set the Dropout rate to 0.5, as it has produced a significant reduction in error for values in the range of [0.3, 0.6] [26]. To optimize the model, we utilized the Adaptive Moment Estimation (Adam) optimizer [12] and trained it over 100 epochs.

We implemented the source code to preprocess the distance maps and develop the model architecture using the Python programming language (version 3.7.9), along with consolidated machine learning libraries and neural networks such as TensorFlow [1] and Keras [4].

2.5   Experimental Design

We split the dataset into training and testing subsets, with 80% and 20% of the data, respectively. A test set was previously extracted by randomly selecting samples from the initial dataset. No data augmentation was applied to test set. For the training set, we used 80% of the samples for tuning the hyperparameters of the network, while the remaining 20% was reserved for validating the model. Commonly, a percentage $\gamma < 0.5$ of the training data is used to validate the model [8]. We stopped adjusting parameters when the number of training epochs reached a predefined value.

We employ an alternative version of the cross-validation (CV) technique approach known as k-fold CV [24]. This technique randomly divides the training set into k subsets of equal size (n/k), where n is the total number of training samples. In this case, we define $k = 5$ because it's possible to guarantee that $\gamma \geq 0.1$, often recommended [8]. One subset is reserved for validation, and the remaining $k - 1$ subsets are used for parameter estimation. We repeated this process k times rotating the validation subset each time. In the end, We estimated the performance based on the average of the k error rates corresponding to each one of k partitions [8]. Since the problem involves multiclass classification, we selected the categorical cross-entropy loss function to train the model.

In this particular problem, the distribution of classes relative to sequence clusters and sub-datasets of peptides from Propedia is unbalanced. To evaluate the model's performance, we used complementary metrics to the error rate. We compared the performance of multiple classifiers trained with the same dataset and calculated complementary metrics such as precision, recall, and F1-Score, which help in choosing the optimal classifier from a performance perspective [29]. We obtained a multi-class confusion matrix to calculate these metrics. Additionally, we presented the performance of the developed model as a function of what it correctly predicted by class [29]. To calculate model performance metrics, we utilized the open-source Python scikit-learn library.

The models were implemented on Google's virtual environment, Colab, which provides access to a Jupyter Notebook. The hardware used consisted of a dual-core processor with 13.6 GB of RAM and an L3 cache of 40–50 MB. However, to accelerate the process, an NVIDIA A100-SXM4 GPU with 40GB of memory was also used, along with an additional 89.6 GB of available RAM. Figure 1 presents an overview of how our methodology was applied.

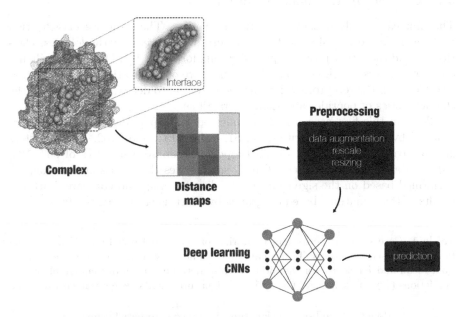

Fig. 1. Overview of the methodology used to evaluate distance map-based signatures. The interface region of the protein-peptide complex is modeled as inter-residue distance maps, used in training the deep neural network. In the end, we want to correctly predict the classes corresponding to the groupings by sequence similarity and peptide type.

3   Results and Discussion

In this study, we use convolutional neural networks to classify interfaces of protein-peptide interactions using computational modeling based on distance maps. We aim to verify if signatures based on contact maps can be as good as graph-based signatures. Graph-based signatures are considered state-of-the-art for classifying macromolecules and, typically, have greater accuracy, in addition to generating vectors of the same size (when the same parameters are applied), allowing direct comparisons without the need for data augmentation steps. Nevertheless, distance map-based signatures can be helpful when combined with convolutional neural networks.

To assess this, we analyzed a problem of high impact in the biotechnology industry: protein-peptide interactions. We collected structures of complexes protein-peptide from the Propedia database. For each collected protein-peptide complex, we extracted the interaction interface. We then computed the distance maps, applied the preprocessing steps, and performed the training and testing process using CNNs (Fig. 1). To evaluate the methodology, we performed two case studies described below.

3.1   Case Study 1: Sequence Clusters

The first case study considers peptide sequences. The sequence classification problem is well established in the literature, with several methodologies, algorithms, and tools that provide optimal results for clustering. On the other hand, our method uses the three-dimensional structure of the peptide and the contact interface with the receptor, which has the potential to detect more details of the characteristics of peptides and their interactions.

Thus, in this first case study, we collected 1,111 peptides from five clusters grouped by sequence similarities: S0, S1, S112, S151, and S162. These clusters are the five most populated in the Propedia database, with sizes of 503, 184, 161, 142, and 122, respectively. Table 1 summarizes the results of the grouping performed based on the signatures of the distance map and compares with the results of the signatures based on graphs by Martins et al. (2023) [19].

Table 1. Case study 1: performance metrics for the model based on CNNs and distance map-based signatures. To evaluate model quality, we calculated accuracy, precision, recall and F1-Score. Furthermore, we also determined the percentage of correct predictions (%) for each class. Graph-based signature results were obtained from [19].

Model	Graph-based signature	Distance-map-based signature
Accuracy	0.92	0.91
Precision	0.92	0.92
Recall	0.92	0.92
F1	0.91	0.91
S0	100%	97.0%
S1	100%	88.0%
S112	100%	93.0%
S151	85.3%	82.0%
S162	40.5%	95.0%

As we hypothesized, representations based on distance maps proved efficient when we analyzed the similarity of the sequences. From Table 1, we can observe that the predictions related to case study 1 had an accuracy of 0.91, precision of 0.92, recall of 0.92, and F1-score of 0.91. This is comparable to the performance achieved by neural networks trained from state-of-the-art representations such as structural signatures [19] that obtained 0.92, 0.92, 0.92, and 0.91 for accuracy, precision, recall, and F1, respectively.

In the work of Martins et al. [19], the clustering of S0, S1, and S112 groups obtained 100% accuracy. On the other hand, they only obtained 85.3% and only 40.5% for S151 and S162 clusters. Although we obtained a slightly lower result for most of the five groups (97%, 88%, 93%, 82%, and 95%, respectively), our methodology can better handle the classification of the latter group (S162). As a disclaimer, we can argue that an improvement in the parameterization performed in work by Martins et al. [19] could get better results for this last cluster. Still, our primary goal here is to demonstrate that signatures based on distance maps can be as good as signatures based on graphs for classifying protein-peptide interfaces.

Moreover, our analysis showed that the percentage of accurate predictions per class was above 80%, indicating that our approach effectively discriminates between different data classes. These findings highlight the potential of distance maps for sequence analysis, suggesting that they may be particularly useful in scenarios where methods based on structural signatures are not feasible or appropriate.

3.2 Case Study 2: Peptide Types

In the second case study, we consider the role of the type of peptide interacting with the receptor. For example, we look at peptide-protein complexes classified as antimicrobial, enzyme, hormone, membrane, plant, and viral. These classes were obtained from the PDB descriptions and are assigned based on the locals where the peptides were obtained.

Thus, in this second case study, we collected 6,238 contact interface structures from Propedia: a protein-peptide database. We analyzed six Propedia datasets: AntimicrobialDB ($n = 10$), ViralDB ($n = 294$), EnzymeDB ($n = 5,344$), MembraneDB ($n = 152$), HormoneDB ($n = 212$), and PlantDB ($n = 256$). Then, we compared our results with the neural network and graph-based signature experiments described in Martins et al. (2023). Table 2 summarizes the results of the grouping performed based on the signatures of the distance map and compares with the results of the signatures based on graphs by Martins et al. (2023) [19].

In Case Study 2, we observed a decrease in model performance for accuracy, precision, recall, and F1 values compared to Case Study 1. A possible reason for the drop in performance could be the impact of dataset imbalance on the final prediction, leading to a bias toward correctly classifying the cluster with more samples (Enzyme). On the other hand, the Antimicrobial class had low accuracies in both studies, mainly due to the low number of instances ($n = 10$). Despite this, we were able to achieve an accuracy of 0.76. Data imbalance is a challenge in machine learning tasks. For this reason, we have a future perspective to research ways to deal with the imbalance in the datasets used in this case study to improve the accuracy in less populated classes without negatively impacting the most populous.

Also, when we examined the percentages of correct predictions, we can see that we obtained values higher than the structural signatures for the Hormone (86%), Membrane (72%), and Viral (55%) classes. This indicates the model

Table 2. Case study 2: performance metrics for the model based on CNNs and distance map-based signatures. To evaluate model quality, we calculated accuracy, precision, recall and F1-Score. Furthermore, we also determined the percentage of correct predictions (%) for each class. Graph-based signature results were obtained from [19].

Model	Graph-based signature	Distance-map-based signature
Accuracy	0.91	0.76
Precision	0.90	0.80
Recall	0.91	0.76
F1	0.90	0.77
Antimicrobial	22.2%	0.0%
Enzyme	97.5%	91.0%
Hormone	63.5%	86.0%
Membrane	36.3%	72.0%
Plant	79.2%	52.0%
Viral	29.9%	55.0%

successfully classified peptides within subsets corresponding to their function. This success aligns with the sequence-structure-function paradigm. Since distance maps serve as an alternative 2D representation of the three-dimensional structure, the model is expected to accurately classify the peptides in the relevant functional subsets of the Propedia.

4   Conclusion and Perspectives

In conclusion, graph-based methods were slightly superior in almost all comparisons performed. However, distance map-based signature methods obtained close results in the sequence-based classification. Also, in the second case study, they achieved better results for some classes, such as classifying hormones, membranes, and viral proteins. These results shed light on the potential use of distance maps for classifying protein-peptide interface, which could be better explored in future experiments using protein-peptide complexes or other macromolecule complexes.

Data Availability Supplementary material, data, and scripts are available at https://github.com/LBS-UFMG/cnn-distance-maps.

Acknowledgements. The authors thank the funding agencies: CAPES, CNPq, and FAPEMIG.

References

1. Abadi, M., et al.: TensorFlow: large-scale machine learning on heterogeneous distributed systems, pp. 1–16 (2016). arXiv:1603.04467
2. Angelova, A., Drechsler, M., Garamus, V.M., Angelov, B.: Pep-lipid cubosomes and vesicles compartmentalized by micelles from self-assembly of multiple neuroprotective building blocks including a large peptide hormone PACAP-DHA. ChemNanoMat 5(11), 1381–1389 (2019). https://doi.org/10.1002/cnma.201900468
3. Bengio, Y., Courville, A., Vincent, P.: Representation learning: a review and new perspectives. IEEE Trans. Pattern Anal. Mach. Intell. 35(8), 1798–1828 (2013). https://doi.org/10.1109/TPAMI.2013.50
4. Chollet, F.: Deep Learning with Python, 4th edn. Manning, New York (2021)
5. Das, A.A., Sharma, O.P., Kumar, M.S., Krishna, R., Mathur, P.P.: PepBind: a comprehensive database and computational tool for analysis of protein-peptide interactions. Genom. Proteom. Bioinform. 11(4), 241–246 (2013). https://doi.org/10.1016/j.gpb.2013.03.002
6. Defresne, M., Sophie, B., Thomas, S.: Protein design with deep learning. Int. J. Mol. Sci. 22, 1741 (2021)
7. Demšar, J., et al.: Orange: data mining toolbox in Python. J. Mach. Learn. Res. 14(1), 2349–2353 (2013). https://doi.org/10.5555/2567709.2567736
8. Duda, R., Hart, P., Stork, G.: Pattern Classification, 2nd edn. Wiley, New York (2001)
9. Goodfellow, I., Bengio, Y., Courville, A.: Deep Learning. Adaptive Computation And Machine Learning, MIT Press, Cambridge (2016)
10. Ioffe, S., Szegedy, C.: Batch Normalization: accelerating deep network training by reducing internal covariate shift. In: Bach, F., Blei, D. (eds.) Proceedings of the 32nd International Conference on International Conference on Machine Learning, vol. 37, pp. 448–456 (2015). arXiv:1502.03167
11. Iyer, M., Jaroszewski, L., Sedova, M., Godzik, A.: What the protein data bank tells us about the evolutionary conservation of protein conformational diversity. Protein Sci. 31, e4325 (2022). https://doi.org/10.1002/pro.4325
12. Kingma, D., Ba, J.: Adam: a method for stochastic optimization. In: Bengio, Y., LeCun, Y. (eds.) Proceedings of the 3rd International Conference on Learning Representations, ICLR 2015 (2015). arXiv:1412.6980
13. Kloczkowski, A., et al.: Distance matrix-based approach to protein structure prediction. J. Struct. Funct. Genom. 10(1), 67–81 (2009). https://doi.org/10.1007/s10969-009-9062-2
14. Lau, J.L., Dunn, M.K.: Therapeutic peptides: historical perspectives, current development trends, and future directions. Bioorg. Med. Chem. 26(10), 2700–2707 (2018). https://doi.org/10.1016/j.bmc.2017.06.052
15. LeCun, Y., et al.: Backpropagation applied to handwritten zip code recognition. Neural Comput. 1(4), 541–551 (1989). https://doi.org/10.1162/neco.1989.1.4.541
16. London, N., Movshovitz-Attias, D., Schueler-Furman, O.: The structural basis of peptide-protein binding strategies. Structure 18(2), 188–199 (2010). https://doi.org/10.1016/j.str.2009.11.012
17. Mariano, D., et al.: A computational method to propose mutations in enzymes based on structural signature variation (SSV). Int. J. Mol. Sci. 20(2), 333 (2019). https://doi.org/10.3390/ijms20020333
18. Martins, P.M., Santos, L.H., Mariano, D., et al.: Propedia: a database for protein-peptide identification based on a hybrid clustering algorithm. BMC Bioinform. 22, 1 (2021). https://doi.org/10.1186/s12859-020-03881-z

19. Martins, P., et al.: Propedia v2.3: a novel representation approach for the peptide-protein interaction database using graph-based structural signatures. Front. Bioinform. **3**, 1103103 (2023). https://doi.org/10.3389/fbinf.2023.1103103

20. Melo, R.C., et al.: Finding protein-protein interaction patterns by contact map matching. Genet. Mol. Res. **6**(4), 946–963 (2007)

21. Min, S., Lee, B., Yoon, S.: Deep learning in bioinformatics. Brief. Bioinform. **18**(5), 851–869 (2017). https://doi.org/10.1093/bib/bbw068

22. Mishkin, D., Sergievskiy, N., Matas, J.: Systematic evaluation of convolution neural network advances on the ImageNet. Comput. Vis. Image Underst. **161**, 11–19 (2017). https://doi.org/10.1016/j.cviu.2017.05.007

23. Moreno-Camacho, C.A., Montoya-Torres, J.R., Jaegler, A., Gondran, N.: Sustainability metrics for real case applications of the supply chain network design problem: a systematic literature review. J. Clean. Prod. **231**, 600–618. https://doi.org/10.1016/j.jclepro.2019.05.278

24. Mosteller, F., Tukey, J.: Data analysis, including statistics. In: Lindzey, G., Aronson, E. (eds.) Revised Handbook of Social Psychology, vol. 2, pp. 80–203 (1968)

25. Pires, D.E.V., de Melo-Minardi, R.C., da Silveira, C.H., Campos, F.F., Meira, W.: aCSM: noise-free graph-based signatures to large-scale receptor-based ligand prediction. Bioinformatics **29**(7), 855–861 (2013). https://doi.org/10.1093/bioinformatics/btt058

26. Srivastava, N., Hinton, G., Krizhevsky, A., Sutskever, I., Salakhutdinov, R.: Dropout: a simple way to prevent neural networks from overfitting. J. Mach. Learn. Res. **15**(56), 1929–1958 (2014). https://doi.org/10.5555/2627435.2670313

27. Theodoridis, S., Koutroumbas, K.: Pattern Recognition, 2nd edn. Academic Press, Burlington (2009)

28. Vinogradov, A.A., Yin, Y., Suga, H.: Macrocyclic peptides as drug candidates: recent progress and remaining challenges. J. Am. Chem. Soc. **141**(10), 4167–4181 (2019). https://doi.org/10.1021/jacs.8b13178

29. Webb, A., Copsey, K.: Statistical Pattern Recognition. Wiley, New York (2011)

30. Xu, M., Yoon, S., Fuentes, A., Park, D.S.: A Comprehensive survey of image augmentation techniques for deep learning. Pattern Recogn. **137**, 109347 (2023). https://doi.org/10.1016/j.patcog.2023.109347

A Power Law Semantic Similarity from Gene Ontology

Eric Augusto Ito[1], Fábio Fernandes da Rocha Vicente[1],
Luiz Filipe Protasio Pereira[2], and Fabricio Martins Lopes[1]([✉])

[1] Computer Science Department, Universidade Tecnológica Federal do Paraná
(UTFPR), Alberto Carazzai, 1640, 86300-000 Cornélio Procópio, PR, Brazil
`fabricio@utfpr.edu.br`
[2] Empresa Brasileira de Pesquisa Agropecuária,
Embrapa Café, Braília DF, 70770-901, Brazil

Abstract. Currently, there is a massive data generation in the most diverse areas of knowledge, as bioinformatics that generates huge amounts of data, requiring the analysis and the summarization of this data for its understanding. Semantic similarity can be seen as an approach that considers the features of objects in a context in order to establish the similarity or dissimilarity of these objects. The Gene Ontology (GO) has been widely employed as a source of features in the estimation of semantic similarity between its terms. Several methods have been proposed in the literature for estimating semantic similarity from GO. However, the methods are based on parametric distributions or arbitrarily defined parameters that do not consider the distribution of GO data. In this context, this work presents a data-driven method for estimating the semantic similarity from GO terms that exploit the power-law distribution. A set of five metabolic pathways were considered for the evaluation of the proposed method and compared with some of the principal methods in the literature. The results showed the adequacy of the proposed method in the estimation of semantic similarities and that it produced more compact gene clusters among all the methods adopted and with an adequate distance between them, leading to clusters more assertive and less susceptible to errors. The proposed method is freely available at https://github.com/EricIto/plawss.

Keywords: Semantic similarity · Complex networks · Power-law · Bioinformatics · Pattern Recognition

1 Introduction

The evolution of technologies has allowed the generation of large amounts of data in various areas of knowledge. As a consequence, new methodologies are being developed with the objective of analyzing and extracting information from this large set of constantly updated data and contribute to the generation of knowledge derived from this data.

M. S. Reis and R. C. de Melo-Minardi (Eds.): BSB 2023, LNBI 13954, pp. 123–135, 2023.
https://doi.org/10.1007/978-3-031-42715-2_12

Semantic similarity (SS) is a fundamental concept, which can be seen as an approach to compare objects from their features [22]. This technique is widely applied in many areas of knowledge, such as information retrieval, biomedicine and artificial intelligence [1,15,28] to cite but a few.

The gene ontology (GO) [11] has been successfully used in many SS applications, such as gene clustering [10], prediction of protein function [19] and validation of gene-gene interactions [8]. Briefly, GO [11] is composed of directed acyclic graphs (DAGs) to define the knowledge about a gene considering three ontologies: molecular function (MFO), biological process (BPO) and cellular component (CCO). Each node is a GO term and two GO terms are connected by different types of edges indicating different relationships. Therefore, obtaining the semantic similarity between GO terms is essential in bioinformatics research, since it represents the relationships between genes based on their annotations.

In recent years many works has been proposed to infer the similarity between pairs of genes or sets of multiple genes [27]. Among the proposed methods for the inference of semantic similarity, two strategies stand out. The first one is based on obtaining the nodes of the graph to obtain the information content (IC) [31]. On the other hand, another strategy is to rely on the edges of the graph to analyze its topology [36]. Thus, these two approaches have been integrated into hybrid methodologies that aim to obtain the advantages of both approaches, and as a consequence, produce more suitable results [34,38]

A hybrid approach is presented in the GOGO [38], which is based on the method proposed by Song et al. [34] and also considers the IC approach indirectly. The GOGO method describes that the IC of the term GO has a high correlation with the number of children of the term. In this way, the method weighs the semantic relations between the GO terms considering the number of children, and thus considering more information to calculate the SS between gene pairs. However, the GOGO method is defined by considering two constant parameters, c and d, to measure the semantic weight. Since the GO terms and the organism annotations are constantly updated, constant values can quickly become out of date.

In this context, this work presents a data-driven methodology for calculating semantic similarity based on gene ontology. More specifically, an alternative SS method is proposed based on the distribution of the number of children of the GO terms as a rule for penalizing its specificity. In addition, the distribution of the number of children per node can be approximated to a power law [5], making the proposed method guided by the ontology distribution and, as a result of this adjustment, the results have a more suitable SS distribution.

2   GO Semantic Similarity

Gene ontology (GO) [11] is an essential source of data for the functional analysis of genes. GO provides an unified vocabulary that describes the functions of genes and relates them considering three ontologies: Biological process (BPO), Molecular Functions (MFO) and Cell Component (CCO). BPO represents a series of molecular functions, which refer to a biological objective that a gene or its gene product contributes. The MFO describes the biochemical activities of a gene product and the CCO describes the location of the occurrence of a molecular function.

Each of these ontologies forms an acyclic graph, each term being represented by a node while the edge represents the relationship between the nodes. This relationship can be of several types, among the most common are: "is_a", "part_of" and "regulates".

Thus, when a gene is investigated and its participation in a biological process, molecular function or location is discovered, that gene receives one or more terms from the respective ontologies in which its participation is showed. Therefore, the gene annotations and the ontology are constantly being updated.

3 Complex Network

Complex networks have been successfully applied to analyze, represent and understand complex systems in many application areas [6,7,12,17,21,24,26,35], leading to a truly multidisciplinary contribution. In particular, networks with a power-law degree distributions, called scale-free networks, have been attracted great attention in the literature [2,33].

The scale-free networks [5] do not have a homogeneous distribution of k connections between their nodes, presenting few vertices highly connected (hubs) to other network nodes, and a large number of nodes with few connections [12]. More specifically, the probability $P(k)$ of a network node to interact with k other nodes decays as a power-law, defined in Eq. 1.

$$P(k) \sim k^{-\gamma}, \tag{1}$$

where γ define the exponential decay, implying an irregular distribution among the network nodes. In fact, the degree distribution is an important complex network property to represent the topological organization of real networks [3, 12].

Regarding biological networks, the scale-free model [5] proved to be adequate to represent metabolic, protein, and gene interaction networks [2–4,18,20,23,30] even considering different organisms. In fact, the scale-free networks present some interesting properties that can led to important biological insigths. For instance, the existence of hubs can provide robustness to random disruptions in biological networks [16] and robustness against perturbations [2], evidence of hierarchical modularity in metabolic networks [30], genomic duplication-divergence of proteins [13], to cite but a few.

In addition, many biological networks are inherently modular, i.e., their functionality can be partitioned into smaller systems or components [25]. In fact, various biological systems present a hierarchical modularity in which the same structures of modules occur repeatedly at different hierarchical levels or topological scales of the network [32]. Thus, the hierarchical modularity and the scale-free are two important properties present in biological networks [3,30].

4   Materials and Methods

4.1   Materials

In order to evaluate the proposed method, a dataset composed of five pathways of *Saccharomyces cerevisiae* was adopted, namely: mevalonete, mannose degradation, phenylalanine degradation, valine degradation and superoxide radicals degradation. Table 1 presents an overview of the clusters formed by genes of the adopted pathways.

Table 1. Clusters formed by the genes of the following metabolic pathways: superoxide radical degradation, mevalonate, mannose degradation, phenylalanine degradation and valine degradation of the SGD database [9].

Pathways	Clusters
superoxide radicals degradation	Cluster 1: [*SOD1,SOD2*]
	Cluster 2: [*CTT1,CTA1*]
mevalonate pathway	Cluster 1: [*ERG10,ERG13*]
	Cluster 2: [*HMG1,HMG2*]
	Cluster 3: [*ERG12,ERG8*]
	Cluster 4: [*MVD1,IDI1*]
mannose degradation	Cluster 1: [*GLK1,HXK1,HXK2*]
	Cluster 2: [*PMI40*]
phenylalanine degradation	Cluster 1: [*ARO8,ARO9*]
	Cluster 2: [*ARO10,PDC1,PDC5,PDC6*]
	Cluster 3: [*SFA1,ADH1,ADH2,ADH3,ADH4,ADH5*]
valine degradation	Cluster 1: [*BAT1,BAT2*]
	Cluster 2: [*PDC1,PDC5,PDC6*]
	Cluster 3: [*SFA1,ADH4,ADH5*]

More specifically, the mevalonate pathway is found in animals, fungi, the cytoplasm of phototrophic organisms, archaea and some eubacteria. The mevalonate pathway is a source of isopentenyl diphosphate in all living organisms. Mannose degradation pathway is a fermentable six-carbon monosaccharide that can be utilized for carbon and energy in *Saccharomyces cerevisiae*. It is also required for many important mannosylation reactions in the cell. Phenylalanine degradation pathway has as result to be a source of nitrogen. Valine degradation is a path to carbon source using amino acid. The superoxide radicals degradation pathway is a defence mechanism of organisms living in an aerobic environment to deal with oxidative stress. These metabolic pathways were adopted, because they are commonly used as benchmarks by competitor methods [34,36,38]. Thus, the evaluation is straightforward and can be extended to other methods available in the literature.

The genes contained in the adopted dataset, in particular each of the pathways, as well as the relationships between genes and GO terms were obtained from the *Saccharomyces* Genome Database (SGD) [9].

4.2   Methods

The proposed method addresses a new perception to infer the semantic similarity (SS) between GO terms. Unlike other works that do not take into account the distribution of data and adopt constants, this study proposes a data-driven method that provide a simplified representation of GO hierarquical structure. More specifically, it is based on analysing the data itself, without having to make any assumptions about parameters, which are arbitrarily defined for other methods [36, 38].

The GO reports the current state of knowledge in biology considering three ontologies: cellular component (CCO), biological process (BPO), and molecular function (MFO), each one represented by a root ontology term. The structure of GO can be described in terms of a graph, where each GO term is a node, and the relationships between the terms are edges between the nodes. GO is hierarchical, with 'child' terms being more specialized than their 'parent' terms.

The proposed methodology is based on a power distribution to penalize the connections between the terms GO, since the greater the number of children that the parent node has, the less specific is the connection with its children. This choice is based on the analysis of the distribution of children contained in the ontology. Figures 1a and 1b show the respective log-log distributions of the number of descendants with the relations "is_a" and "part_of" of GO terms and a power-law distribution, which indicate that the distribution of the number of children of the GO terms is approximately a power-law.

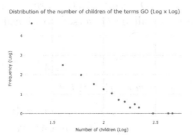
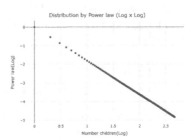
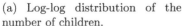

(a) Log-log distribution of the number of children.

(b) Log-log distribution of data by power law.

Fig. 1. Log-log function of the number of children in Gene Ontology compared to the data distribution described by the power law.

It is important to confirm that the distributions of the number of children of the ontology and the power law distribution are similar, i.e. they are comparable to each other. Therefore, an experiment was performed in order to evaluate these distributions. First was obtained the average among the number of children from GO, achieving 1.81, which was considered for the power law exponent.

Thus, the histograms with the probability distributions of: (i) the power law and (ii) the number of children per GO term were considered. The two-sample Kolmogorov-Smirnov test [29] was performed, which is a non-parametric method for comparing two samples. Thus, the two histograms of probability were considered, in order to verify whether they were statistically different. The p-value obtained was 0.586, thus being greater than 0.05 of significance, indicating that is not possible to reject the null hypothesis that the values of the probability histograms are similar.

Semantic Similarity Between GO Terms. To analyze the similarity between a pair of GO terms, a graph is built for each GO term that contain all of its predecessors up to the root of the gene ontology. This step is performed separately for each of the ontologies: biological process, molecular functions, and cellular component. With the graphs generated, the first step is to calculate the weight that each GO term in relation to the information content it carries. In this step the power law is adopted to define the weight (w_c) taking into account how many children the parent of the term has and also what kind of relation to the parent term. In this work, the definition of previous works regarding the semantic weight of relations is adopted, assigning a value of 0.8 for the "is_a" relation and 0.6 for the "part_of" relation [36]. Equation 2 defines the semantic weight of each GO term.

$$w_c = k^{(w_s - 1)} \tag{2}$$

where k is the number of children that parent of the GO term has and w_s is the semantic weight.

For each GO term A is defined a semantic value (S_A) that measures its semantic value, this value depends on the weight (w_c), as well as the semantic value of its descendants, as defined by Eq. 3. GO terms, if compared to itself receive value 1, otherwise the path to the root will be traversed to identify its semantic value.

$$\begin{cases} S_A(t) = 1 & \text{if } t = A \\ S_A(t) = Max\{w_c \times S_A(t') | t' \in children(t)\} & \text{if } t \neq A \end{cases} \tag{3}$$

The semantic similarity between GO terms $S_{GO}(A, B)$ is defined by dividing their common ancestor semantic values by the sum of all the semantic values of the GO terms, as defined by the Eq. 4.

$$S_{GO}(A, B) = \frac{\sum\limits_{t \in T_A \cap T_B} (S_A(t) + S_B(t))}{\sum\limits_{t \in T_A} S_A(t) + \sum\limits_{t \in T_B} S_B(t)} \tag{4}$$

Thus, the semantic content of each GO term is considered by taking into account the information content of all of its ancestor terms in the GO graph, i.e. the targets of term A (T_A) and term B (T_B). The overview of the proposed methodology for GO term similarity is shown in Fig. 2.

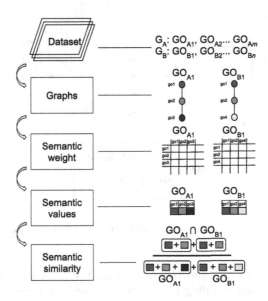

Fig. 2. Overview of the proposed methodology for GO term similarity.

Semantic Similarity Between Genes. The semantic similarity between pairs of genes G_A and G_B is performed by the combination of each GO term of Gene A (GO_A), with all GO terms of Gene B (GO_B) and vice-versa, taking into account the highest similarity value of each combination, as defined by the Eq. 5. Next, a clustering of the similarity values between the GO terms is performed by considering the Affinity Propagation algorithm [14]. Therefore, the semantic similarity between a pair of genes is defined by the maximum of the average similarity values of the generated clusters.

$$Sim(G_A, G_B) = \max_{GO_A \in G_A GO_B \in G_B} S_{GO}(GO_A, GO_B) \qquad (5)$$

Figure 3 shows the overview of the proposed methodology for the semantic similarity between genes.

5   Results and Discussion

In order to assess the performance of the proposed method PLAWSS, as well as to compare its results, some of the principal semantic similarity methods from GO were considered: GOGO [38], Wang [36] and Resnik [31]. Although the proposed method is suitable for the estimation of semantic similarity in any of the three gene ontologies: cellular location, molecular function and biological process, to perform the experiments, the biological process ontology was considered, which is commonly adopted by competing methods.

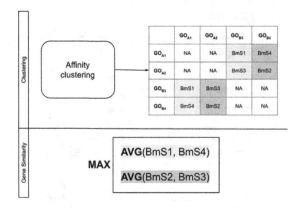

Fig. 3. Calculation of semantic similarity between genes.

The results were evaluated by considering the semantic similarity of the expected clusters corresponding to groups of genes that have the same or very similar functions in each pathway, as presented in Sect. 4.1. An example is the superoxide radicals degradation, in which one cluster is formed by genes *SOD1* and *SOD2* and another cluster is formed by genes *CTT1* and *CTA1*. Besides the superoxide radicals degradation pathway, another four pathways were adopted to assess the semantic similarity methods.

Table 2 presents the semantic similarities of the proposed method for the genes in the superoxide radicals degradation pathway, for each gene pair present in the pathway the semantic similarity is performed, it is expected that genes that belong to a cluster have higher similarities and genes in different cluster have lower semantic similarities.

It can be noted that the proposed method produced high semantic similarity values for genes in the same cluster, highlighted in bold, while it produced lower similarity values for genes from different clusters. The competing methods produced lower similarity values for genes in the same cluster than the proposed method. The GOGO and Wang methods produced slightly lower similarity values for genes from different clusters, while the Resnik method produced a high similarity value between *CTA1* and *SOD2* genes which belong to different clusters.

In order to analyse the overall efficiency of each method considering the adopted metabolic pathways and the respective generated clusters, two indexes commonly used in cluster analysis and pattern recognition were adopted: Complete Diameter Distance (CDD) and Average Centroid Linkage Distance (ACLD) [37]. In this way, CCD represents the distance between the most remote samples belonging to the same cluster, thus it is possible to evaluate how compact the generated clusters are. On the other hand, ACLD reflects the average distance between the centroids of generated clusters, making it possible to assess the distance between the clusters.

Table 2. Semantic similarity matrix of the adopted methods for the genes in superoxide radicals degradation pathway.

		SOD1	*SOD2*	*CTT1*	*CTA1*
PLAWSS	*SOD1*	**1.00**	**0.78**	0.31	0.44
	SOD2	**0.78**	**1.00**	0.31	0.53
	CTT1	0.31	0.31	**1.00**	**0.82**
	CTA1	0.44	0.53	**0.82**	**1.00**
GOGO	*SOD1*	**1.00**	**0.39**	0.10	0.21
	SOD2	**0.39**	**1.00**	0.17	0.45
	CTT1	0.10	0.17	**1.00**	**0.45**
	CTA1	0.21	0.45	**0.55**	**1.00**
Wang	*SOD1*	**1.00**	**0.49**	0.17	0.31
	SOD2	**0.49**	**1.00**	0.29	0.51
	CTT1	0.17	0.29	**1.00**	**0.61**
	CTA1	0.31	0.51	**0.61**	**1.00**
Resnik	*SOD1*	**1.00**	**0.41**	0.21	0.36
	SOD2	**0.41**	**1.00**	0.36	0.71
	CTT1	0.21	0.36	**1.00**	**0.45**
	CTA1	0.36	0.71	**0.45**	**1.00**

All adopted metabolic pathways were evaluated and the semantic similarity matrices produced. Therefore, the evaluation indexes (CDD and ACLD) of the clusters produced for each metabolic pathway were performed and can be seen in Tables 3 and 4.

Table 3. Complete Diameter Distance (CDD) of the adopted methods for the clusters in mevalonate, mannose degradation, phenylalanine degradation and valine degradation pathways.

Metabolic pathway	PLAWSS	GOGO	WANG	RESNIK
superoxide radicals degradation	**0.40**	1.07	0.90	1.15
mevalonate pathway	0.79	1.21	**0.74**	1.58
mannose degradation	**0.00**	0.31	0.24	0.49
phenylalanine degradation	**0.74**	0.94	0.75	1.17
valine degradation	**0.37**	0.69	0.52	1.22
Average	**0.46**	0.84	0.63	1.12

Regarding the CDD, it is possible to verify that the proposed method produces more compact clusters for the metabolic pathways mannose degradation, phenylalanine degradation and valine degradation, with highlight to mannose degradation with a value of 0, showing that all the genes of each cluster have semantic similarity equal 1 (maximum) among them.

Table 4. Average Centroid Linkage Distance (ACLD) of the adopted methods for the clusters in mevalonate, mannose degradation, phenylalanine degradation and valine degradation pathways.

Metabolic pathway	PLAWSS	GOGO	WANG	RESNIK
superoxide radicals degradation	**0.50**	**0.50**	0.45	0.31
mevalonate pathway	0.29	**0.30**	0.08	0.18
mannose degradation	**0.84**	0.76	0.67	0.47
phenylalanine degradation	0.45	**0.46**	0.33	0.28
valine degradation	0.49	**0.51**	0.34	0.31
Average	**0.52**	0.51	0.37	0.31

Regarding the ACLD, it is possible to verify that the proposed method produces clusters that are distant from each other, specially for the mannose degradation pathway, with the best result among all the methods, while in the other pathways it presented results equivalent to the GOGO method.

In summary, considering the average values among the evaluated pathways, the proposed method showed an average CDD value of 0.46, while the other methods GOGO (0.84), Wang (0.63) and Resnik (1.12). On the other hand, considering the mean values of ACLD, the proposed method present 0.52, GOGO (0.51), Wang (0.37) and Resnik (0.31).

The competing methods produced cluster less compact and closer to each other, and therefore more susceptible to noise and clustering errors. Therefore, it was possible to identify that the proposed method produced the best balance with more compact clusters and with adequate spacing, and therefore more assertive and less susceptible to noise and clustering errors.

6   Conclusion

Approaches to measure the similarity and semantic relationship between terms can provide semantic context needed for the identification and characterisation of relationships, with special emphasis on problems in bioinformatics, which present a large amount of data to be analysed and better understood.

Currently, there are several methods that address this context, but without considering the distribution of the data, which leads to imposing arbitrarily defined parameters and the use of parametric distributions. This work presents a new data-driven method to perform semantic similarity from GO.

Five metabolic pathways of *Saccharomyces cerevisiae* commonly used by similar methods in the literature were adopted for evaluation of the proposed method. Some of the principal methods of semantic similarity estimation were considered for comparison of the results, such as GOGO, Wang and Resnik. The results showed that the proposed method is suitable and functional for semantic similarity estimation in all metabolic pathways adopted. In addition, the generated clusters were analysed in terms of indices for the evaluation of compactness and distance between the generated clusters.

The proposed method produced, on average, the most compact clusters among all methods and with a suitable distance between them, leading to more adequate and assertive clusters and less susceptible to clustering errors.

Acknowledgment. This study was funded by the Coordenação de Aperfeiçoamento de Pessoal de Nível Superior (CAPES), Conselho Nacional de Desenvolvimento Científico e Tecnológico (CNPq) (grant 440412/2022-6) and the Fundação Araucária and SETI (grant 138/2021 and NAPI - Bioinformática - grant PDI 66/2021).

References

1. Akmal, S., Shih, L.H., Batres, R.: Ontology-based similarity for product information retrieval. Comput. Ind. **65**(1), 91–107 (2014)
2. Albert, R.: Scale-free networks in cell biology. J. Cell Sci. **118**(21), 4947–4957 (2005)
3. Almaas, E., Barabási, A.L.: Power Laws in Biological Networks. Springer, Boston (2006). https://doi.org/10.1007/0-387-33916-7_1
4. Barabási, A.L.: Scale-free networks: a decade and beyond. Science **325**(5939), 412–413 (2009)
5. Barabási, A.L., Albert, R.: Emergence of scaling in random networks. Science **286**(5439), 509–512 (1999)
6. Barabási, A.L., Gulbahce, N., Loscalzo, J.: Network medicine: a network-based approach to human disease. Nat. Rev. Genet. **12**(1), 56–68 (2011)
7. Boccaletti, S., Latora, V., Moreno, Y., Chavez, M., Hwang, D.U.: Complex networks: structure and dynamics. Phys. Rep. **424**(4–5), 175–308 (2006)
8. Cao, R., Cheng, J.: Deciphering the association between gene function and spatial gene-gene interactions in 3d human genome conformation. BMC Genom. **16**(1), 880 (2015)
9. Cherry, J.M., et al.: SGD: saccharomyces genome database. Nucleic Acids Res. **26**(1), 73–79 (1998)
10. Cho, Y.R., Zhang, A., Xu, X.: Semantic similarity based feature extraction from microarray expression data. Int. J. Data Min. Bioinform. **3**(3), 333–345 (2009)
11. Gene Ontology Consortium: Expansion of the gene ontology knowledgebase and resources. Nucleic Acids Res. **45**(D1), D331–D338 (2016)
12. Costa, L.F., Rodrigues, F.A., Travieso, G., Villas-Boas, P.R.: Characterization of complex networks: a survey of measurements. Adv. Phys. **56**(1), 167–242 (2007)
13. Evlampiev, K., Isambert, H.: Conservation and topology of protein interaction networks under duplication-divergence evolution. Proc. Natl. Acad. Sci. **105**(29), 9863–9868 (2008)
14. Frey, B.J., Dueck, D.: Clustering by passing messages between data points. Science **315**(5814), 972–976 (2007)
15. Garla, V.N., Brandt, C.: Semantic similarity in the biomedical domain: an evaluation across knowledge sources. BMC Bioinform. **13**(1), 261 (2012)
16. He, X., Zhang, J.: Why do hubs tend to be essential in protein networks? PLOS Genet. **2**(6), 1–9 (2006)

17. Ito, E.A., Katahira, I., Vicente, F.F., Pereira, L.P., Lopes, F.M.: BASiNET-BiologicAl Sequences NETwork: a case study on coding and non-coding RNAs identification. NAR **46**(16), e96 (2018)

18. Jeong, H., Tombor, B., Albert, R., Oltvai, Z.N., Barabási, A.L.: The large-scale organization of metabolic networks. Nature **407**, 651–654 (2000)

19. Jiang, Y., et al.: An expanded evaluation of protein function prediction methods shows an improvement in accuracy. Genome Biol. **17**(1), 184 (2016)

20. Khanin, R., Wit, E.: How scale-free are biological networks. J. Comput. Biol. **13**(3), 810–818 (2006)

21. de Lima, G.V.L., Castilho, T.R., Bugatti, P.H., Saito, P.T.M., Lopes, F.M.: A complex network-based approach to the analysis and classification of images. In: CIARP 2015. LNCS, vol. 9423, pp. 322–330. Springer, Cham (2015). https://doi.org/10.1007/978-3-319-25751-8_39

22. Lin, D., et al.: An information-theoretic definition of similarity. In: ICML, vol. 98, pp. 296–304 (1998)

23. Lopes, F.M., Martins Jr, D.C., Barrera, Jr., Cesar, Jr., Roberto M.: A feature selection technique for inference of graphs from their known topological properties: revealing scale-free gene regulatory networks. Inf. Sci. **272**, 1–15 (2014)

24. Lopes, F.M., Martins, D.C., Barrera, J., Cesar, R.M.: SFFS-MR: a floating search strategy for GRNs inference. In: Dijkstra, T.M.H., Tsivtsivadze, E., Marchiori, E., Heskes, T. (eds.) PRIB 2010. LNCS, vol. 6282, pp. 407–418. Springer, Heidelberg (2010). https://doi.org/10.1007/978-3-642-16001-1_35

25. Lorenz, D.M., Jeng, A., Deem, M.W.: The emergence of modularity in biological systems. Phys. Life Rev. **8**(2), 129–160 (2011)

26. Newman, M.E.J.: The structure and function of complex networks. SIAM Rev. **45**(2), 167–256 (2003)

27. Pesquita, C.: Semantic similarity in the gene ontology. In: The Gene Ontology Handbook, pp. 161–173. Humana Press, New York, NY (2017)

28. Pesquita, C., Faria, D., Falcao, A.O., Lord, P., Couto, F.M.: Semantic similarity in biomedical ontologies. PLoS Comput. Biol. **5**(7), e1000443 (2009)

29. Pratt, J.W., Gibbons, J.D.: Kolmogorov-Smirnov two-sample tests. In: Pratt, J.W., Gibbons, J.D. (eds.) Concepts of Nonparametric Theory. Springer Series in Statistics, pp. 318–344. Springer, New York, NY (1981). https://doi.org/10.1007/978-1-4612-5931-2_7

30. Ravasz, E.: Detecting Hierarchical Modularity in Biological Networks, pp. 145–160. Humana Press, Totowa, NJ (2009)

31. Resnik, P.: Semantic similarity in a taxonomy: an information-based measure and its application to problems of ambiguity in natural language. J. Artif. Intell. Res. **11**, 95–130 (1999)

32. Serban, M.: Exploring modularity in biological networks. Philos. Trans. R. Soc. B **375**(1796), 20190316 (2020)

33. Shirai, S., et al.: Long-range temporal correlations in scale-free neuromorphic networks. Netw. Neurosci. **4**(2), 432–447 (2020)

34. Song, X., Li, L., Srimani, P.K., Yu, P.S., Wang, J.Z.: Measure the semantic similarity of go terms using aggregate information content. IEEE/ACM Trans. Comput. Biol. Bioinf. **11**(3), 468–476 (2014)

35. da Rocha Vicente, F.F., Lopes, F.M.: SFFS-SW: a feature selection algorithm exploring the small-world properties of GNs. In: Comin, M., Käll, L., Marchiori, E., Ngom, A., Rajapakse, J. (eds.) PRIB 2014. LNCS, vol. 8626, pp. 60–71. Springer, Cham (2014). https://doi.org/10.1007/978-3-319-09192-1_6

36. Wang, J.Z., Du, Z., Payattakool, R., Yu, P.S., Chen, C.F.: A new method to measure the semantic similarity of GO terms. Bioinformatics **23**(10), 1274–1281 (2007)
37. Webb, A.R.: Statistical Pattern Recognition, 2nd edn. John Willey & Sons, New York (2002)
38. Zhao, C., Wang, Z.: GOGO: an improved algorithm to measure the semantic similarity between gene ontology terms. Sci. Rep. **8**(1), 1–10 (2018)

Gene Networks Inference
by Reinforcement Learning

Rodrigo Cesar Bonini[✉] and David Correa Martins-Jr

Centro de Matemática, Computação e Cognição - Universidade Federal do ABC,
Santo, André-SP, Brazil
{rodrigo.bonini,david.martins}@ufabc.edu.br

Abstract. Gene Regulatory Networks inference from gene expression
data is an important problem in systems biology field, involving the
estimation of gene-gene indirect dependencies and the regulatory functions among these interactions to provide a model that explains the gene
expression dataset. The main goal is to comprehend the global molecular
mechanisms underlying diseases for the development of medical treatments and drugs. However, such a problem is considered an open problem, since it is difficult to obtain a satisfactory estimation of the dependencies given a very limited number of samples subject to experimental
noises. Many gene networks inference methods exist in the literature,
where some of them use heuristics or model based algorithms to find
interesting networks that explain the data by codifying whole networks
as solutions. However, in general, these models are slow, not scalable to
real sized networks (thousands of genes), or require many parameters,
the knowledge from an specialist or a large number of samples to be
feasible. Reinforcement Learning is an adaptable goal oriented approach
that does not require large labeled datasets and many parameters; can
give good quality solutions in a feasible execution time; and can work
automatically without the need of a specialist for a long time. Therefore, we here propose a way to adapt Reinforcement Learning to the
Gene Regulatory Networks inference domain in order to get networks
with quality comparable to one achieved by exhaustive search, but in
much smaller execution time. Our experimental evaluation shows that
our proposal is promising in learning and successfully finding good solutions across different tasks automatically in a reasonable time. However,
scalabilty to networks with thousands of genes remains as limitation of
our RL approach due to excessive memory consuming, although we fore-
see some possible improvements that could deal with this limitation in
future versions of our proposed method.

Keywords: reinforcement learning · gene regulatory networks
inference · boolean networks

© The Author(s), under exclusive license to Springer Nature Switzerland AG 2023
M. S. Reis and R. C. de Melo-Minardi (Eds.): BSB 2023, LNBI 13954, pp. 136–147, 2023.
https://doi.org/10.1007/978-3-031-42715-2_13

1   Introduction

Systems Biology (SB) is an interdisciplinary research field that focuses on the study of complex networks of molecular interactions existing in live organisms [27]. The functioning of an organism depends on several metabolic pathways regulated by gene expression networks. The development of techniques such as DNA Microarrays [24], SAGE [29], RNA-Seq [30], and Single-Cell RNA-Seq [12] enabled the expression level (mRNA concentrations) measurement of thousands of genes simultaneously and in several experiments (usually patients, treatments or timepoints). Thus, several methods for analyzing the topology and dynamical evolution of the gene expression levels have been proposed, with the goal of reverse engineering the regulatory control mechanisms [14,33]. Currently, the Gene Regulatory Networks (GRN) inference problem is attracting the attention of many researchers, mainly because of the enormous volume of gene expression data generated for many species and specific conditions. Nevertheless, the GRN inferenceis still an open problem.

A common problem presented by gene expression analysis is the huge number of genes (variables) with just a few dozens of samples (experiments), demanding then the development of statistical and computational methods to alleviate the estimation error committed in the presence of small number of samples and high dimensionality. Other factors that contribute to the difficulty of this task are associated to the large degree of imprecision inherent to the gene expression measurements (noisy data), the large complexity of inter-relationship networks, and lack of prior information about many biological organisms [14,20].

There are two main approaches to model the complex networks of gene interactions: continuous and discrete. The continuous approach uses differential equations to reach a quantitative detailed model of biochemical networks with cellular functions [9]. There is also a discrete approach, where it is based on the construction of qualitative discrete models of gene interactions, including the models based on graphs like Boolean Networks (BNs) [17] and its variation, the Probabilistic Boolean Networks (PBNs) [25]. The continuous approaches can provide a more detailed comprehension of the considered system, but they require a significant number of samples and information about the characteristics of the reactions, which is rare and difficult to have [14,20].

In the context of discrete models, BNs represent an appropriate model to generalize and capture the global behavior of biological systems, especially when the number of samples available is limited and the dimensionality (number of variables) is large [17]. Such model performs a data quantization, which makes the BN model simpler to work and to adapt [11].

Many gene networks inference methods modeled as BN were proposed in the literature, usually performing Exhaustive Search (ES) with a given size of regulators (predictors) set per target gene over the state transition matrix of quantized gene expression data [1,3,18,22]. Some other approaches, use heuristics or model based algorithms to find interesting networks that explain the data by codifying whole networks as solutions [16,23]. But, these methods are not scalable to real-sized networks with thousands of genes and usually require tuning of many parameters, the knowledge from an specialist, or large number of samples to become feasible.

RL [28] is an approach that allows agents to learn autonomously through interactions with an environment. In RL, an action that affects the environment is chosen by the agent, then the agent observes how much that action helped to the task completion through a reward function. An agent can learn how to optimally solves tasks by executing this process multiple times autonomously.

In order to learn with few parameters, fast and automatically, many decision problems such as GRN inference, might be modeled by a Markov Decision Process (MDP) [28], and RL is an extensively used solution for MDPs. Hence, in this work we propose a GRN inference method using RL, consisting in applying RL algorithms per each target gene independently aiming at obtaining the best predictor set (a single, a pair, a triple or as many predictors the target gene can have). To the best of our knowledge, our proposal is novel, since there is no RL method fully adapted to the GRN inference domain.

With the objective of analyzing the results of the proposed method, experiments involving artificial Boolean networks generated by the Erdös-Renyi random complex network model [13] were performed. We compared the proposed RL method with the traditional ES algorithm that searches for pairs or triples of predictors per target. RL shows promise given the quality of the networks achieved, but with much less exploration of the search space, consequently leading to much smaller execution times.

2   Foundations

In this section we introduce the relevant basic concepts and works involving the Boolean Networks (BN) model for Gene Regulatory Networks (GRN) and Reinforcement Learning (RL).

2.1   Boolean Networks Inference

Genes can indirectly interact with each other, mainly by means of the interaction of part of their generated proteins, which can activate or inactivate the gene transcription (expression) of mRNAs responsible for producing proteins. This process is known as a GRN, which can be represented as a directed graph with genes as nodes and interactions as edges, in which a source node promotes or inhibits the activity (expression) level of a target node [6].

A BN is a model proposed for the study of complex systems dynamics, and of GRNs in particular [17]. In a Boolean Network, a set of n Boolean variables is represented by a set $V = \{v_1, v_2, \ldots, v_n\}$ of vertices in a graph, while the other component of a BN is a set of Boolean transition functions $\Phi = \{\phi_1, \phi_2, \ldots, \phi_n\}$, each function corresponding to one vertex [10]. In GRNs modeling, each vertex v_i is associated with a certain gene, thus we will refer to v_i as either a gene or a vertex, indistinctly. Each gene $v_i \in \{0,1\}, i = 1,2,...,n$ represents a binary variable for which its value in the next time instant $t+1$ is completely determined by the values of its k_i predictor genes in the current time instant t. More concretely, these dynamics can be represented by $v_i(t+1) = f_i(v_{1i}(t), v_{2i}(t),...,v_{ki}(t))$, in which $v_{1i}, v_{2i},...,v_{ki}$ represents the k_i predictor (or regulatory) genes that influence the target gene v_i.

Although BN variables present only two possible values, the BNs inference is still considered an open problem due to the well known curse of dimensionality. An approach that deals with this problem with a certain success is the Probabilistic Gene Networks (PGN), which provides some biologically sound simplifications to the inference process, allowing the application of a local feature selection for each target to look for the best gene subset which predicts the behavior of a given target gene [3,15,16,19], by assuming conditional independence, among other simplification assumptions. The general PGN inference process can be described as follows:

1. Take as input a gene expression matrix, the index of the considered target gene, and a criterion function
2. For each considered subset of predictor genes (predictor subset):
   - A table of joint probabilities (or just countings) among the possible instances (0 or 1) of the target gene and the possible instances (all combinations of bits) of the predictor subset
   - A criterion function is applied to the table filled in the previous step and assigned to the considered predictor subset
3. Return the candidate predictor subsets ordered by the adopted criterion function values

The procedure described above is generic and can be described by efficient heuristic greedy algorithms, although they do not guarantee optimality [2,20,21]. On the other hand, the ES is the only search which guarantees optimality [7], but it is a prohibitely costly combinatorial process, only feasible considering pairs or triples of predictor subsets, since the number of genes is usually in the order of thousands. So our proposal is to apply for the first time the RL framework as a sound solution to infer BNs approached as PGNs.

2.2   Reinforcement Learning

The RL framework [28] allows autonomous agents to learn through interactions in the environment. Many sequential decision problems (as the Boolean Networks problems), can be modeled by a Markov Decision Process (MDP) [28], and RL is an extensively used solution for MDPs driven by environment interactions. An MDP is described by the tuple $\langle S, A, T, R \rangle$, where S is the set of environment **states**, A is the set of available **actions**, T is the **transition function**, and R is the **reward function** (the agent does not know T and R). The goal of the agent in an MDP is to learn an optimal **policy** π^* that maps each state to the actions that lead the agent to the highest expected cumulative sum of rewards over the lifetime of the agent.

As the output from the transition and reward functions cannot be predicted in learning problems, the MDP can be solved through interactions with the environment, which can be accomplished, for instance, by the **Q-Learning** algorithm [31]. Q-Learning iteratively learns a *Q-value*, aiming at estimating the cumulative discounted reward associated with each state-action pair: $Q : S \times A \to \mathbb{R}$. At each decision step, Q is updated following: $Q_{t+1}(s_t, a_t) = (1 - \alpha)Q_t(s_t, a_t) + \alpha[r_{t+1} + \gamma\max_a Q_t(s_{t+1}, a)]$. Q-Learning eventually converges to the optimal Q function: $Q^*(s, a) = E\left[\sum_{i=0}^{\infty} \gamma^i r_i\right]$, and Q^* can be used to define an optimal policy as: $\pi^*(s) = \arg\max_a Q^*(s, a)$.

SARSA algorithm is another approach that expands RL to State, Action, Reward, State, Action, and is an on-policy value-based approach. As a form of value iteration, it needs a value update rule [28].

In SARSA, at each decision step, Q is updated following:

$Q_{(s_t, a_t)} = Q(a_t, a_t) + \alpha(r_t + \gamma(Q(s_{t+1}, a_{t+1}) - Q_t(s_t, a_t))$.

The Q-value update rule is mainly what distinguishes SARSA from Q-learning. In SARSA the time difference value is calculated using the current state-action combo and the next state-action combo. This means we need to know the next action our policy takes in order to perform an update step. This makes SARSA an on-policy algorithm as it is updated based on the current choices of our policy.

Q-learning differs from SARSA in its update rule by assuming the use of the optimal policy. The use of the $\max_a$ function over the available actions makes the Q-learning algorithm an off-policy approach. This is because the policy we are updating differs in behavior from the policy we use to explore the world, which uses an exploration parameter ϵ to choose between the best identified action and a random choice of action. In this work we tested both of them in order to check the influence of the situations and parameters aforementioned.

However, it is important to highlight that the standard Q-Learning, SARSA and classical RL techniques might be inefficient in environments with large state spaces and some adaptations and tuning must be done in order to scale and accelerate these approaches depending on the problem we are dealing with.

3   Using RL to Infer GRNs

As the Exhaustive Search (ES) for Gene Regulatory Networks (GRN) inference requires constraining the search space of the possible gene predictor subsets to pairs or triples to be feasible, resulting in optimal subsets from this constrained search space, we here propose a GRN inference method by using RL, in order to make the learning process automatic, in a less constrained search space, faster, and making the use of just a few parameters. It consists in simulating the behaviour of BNs, modeling them as an MDP. The method takes as input a temporal binary gene expression data matrix (genes in rows and timepoints in columns), then selects a target gene and derive a counting table per explored predictor subset from the input matrix. A counting table is a matrix with the rows representing all possible instances (bit strings) of a predictor subset, the columns meaning the two possible values of the target (0 or 1), and each cell (i, j), where i is the decimal corresponding to a given instance (bit string) and $j \in \{0, 1\}$, stores the number of observations of the target being in the state j in a given timepoint and the predictor subset being in a given state, encoded by a bit string whose conversion from binary to decimal is i, in the previous timepoint.

For each explored predictor subset, the RL algorithm reward (R) is based on the sum of the minimum number of occurrences of the target being 0 or 1 per counting table row (predictor subset instance), which is analogous to the number of times a Bayesian classifier misses the correct target values based on the possible instances (states) of the candidate predictor subset. From now on, for the sake of simplicity, we call this sum as "error". So we can apply RL algorithms such as Qlearning and SARSA for each target gene independently aiming at obtaining the best predictor subset for each target gene (an empty subset, a single, a pair, a triple, or as many predictors the target gene can have), i.e., the subset with minimum error. This is possible by assuming conditional independence, a PGN simplification assumption (see Sect. 2.1).

Using RL for inference of GRNs implies in considering all the predictors and target genes as the State Space, where a state S is one target gene plus a set of selected predictors used to construct a counting table of occurrences of 0's and 1's. An action is the choice of this predictor, where all predictors are part of the set of actions A. Here we also included an special action to allow the agent to **stop** and finish the episode and not to get stuck in a maximum or minimum local. The transition function is considered always to work as deterministic, with $T = 1.0$, not giving any chance to perform other than the right action. And finally, the reward R the agent receives is the difference between the error of the last step and the error of the current step.

Algorithm 1 summarizes our approach. It initially requires a list of predictors, target gene and a gene expression data matrix. Then, we set the error as the highest possible. After that, for a *numberOfEpisodes* it repeats the learning process. During the learning process (lines 4-13), the agent selects actions (predictors available) and remove them from the set of possible actions for the next step to avoid selecting more than once in the same episode/run. The counting table error then is calculated, but here we have a special action, which stops and ends the episode when selected, so the agent does not construct all possible counting tables during the episodes. Finally, the algorithm returns the best set of predictors (with the minimum error found) for each target gene.

4   Experimental Results

All experiments were performed in a regular PC desktop with an Intel(R) Core (TM) i7-7700 CPU (3.6GHz), 16GB RAM DDR3.

The methods were executed 10 times per experiment. A given experiment takes as input 10 gene expression data matrices, each one starting from a distinct randomly chosen initial state. Such gene expression data matrices were generated from a randomly generated artificial Boolean network, assuming random topology (Erdös-Renyi complex network model [13]) and average degree (number of predictors per target) equal to 2.

Algorithm 1 RL for GRN Inference

Require: list of possible predictor genes, target genes, and gene expression data
 1: **while** episode < numberOfEpisodes **do**
 2:     state = []
 3:     lastError = ∞
 4:     **while** action ! = stop **do**
 5:         select action *action* using RL algorithm
 6:         state.append(action)
 7:         predictorsByEpisode.append(action)
 8:         remove selected action from predictors to select
 9:         newError = Counting table error of predictors in *state*
10:         reward = lastError - newError
11:         update RL algorithm according to last state, reward, and action
12:         lastError = newError
13:     **end while**
14: **end while**
15: **return** the best set of predictors found

In order to check the efficiency of our proposal, we executed 3 experiments comparing the performance of RL algorithms with the ES: 1) **20 genes and 20 timepoints**, 2) **20 genes and 50 timepoints**, and 3) **50 genes and 50 timepoints**. In all experiments, the agent started in a random initial state, where the set of states S contains all predictors the algorithm can select for each target, and the actions A are the predictors choices or the special action to "stop" and end the episode, where the RL algorithms can choose to stop the episode at any time (randomly). We evaluated error (considered as our reward by the RL algorithms) starting with ∞. The experiments 1 and 2 were performed over 100 episodes and the experiment 3 over 300 episodes. We adopted the following RL parameters: $\alpha = 0.2$, $\gamma = 0.9$ and $\epsilon = 0.2$.

4.1   GRN with 20 Genes and 20 Timepoints

In this experiment, we evaluated our proposal in 10 gene expression data matrices, each one containing 20 genes and 20 time samples (timepoints). Here, the RL algorithms could learn really fast in terms of number of episodes, where the execution time varies according to the choices done by the algorithm. Figure 1 shows that Qlearning performed slightly better than SARSA. The mean error at the end of the experiments was about 0.5 (considering 10 executions on distinct gene expression data matrices). The minimum error obtained was zero for both Exhaustive Search and RL, but RL could find the minimum error after analyzing 508 predictor subsets, while ES had to analyze all triples (combination of 20, 3 to 3 = 1,140 triples).

4.2   GRN with 20 Genes and 50 Timepoints

In this experiment, we evaluated our proposal in 10 gene expression data matrices, each one containing 20 genes and 50 time samples (timepoints). Here, the RL algorithms took a little more episodes to run (where the time varies according to the choices done by the algorithm) compared to the previous experiments, but still needed less than 100 episodes. Figure 2 shows that Qlearning still performed slightly better than SARSA. The mean error at the end of the experiments was about 5.5 (considering 10 executions on distinct gene expression data matrices). The minimum error obtained was 5.0 for both Exhaustive Search and RL, but RL could find the minimum error after analyzing 942 predictor subsets, while ES had to analyze all triples (combination of 20, 3 to 3 = 1,140 triples).

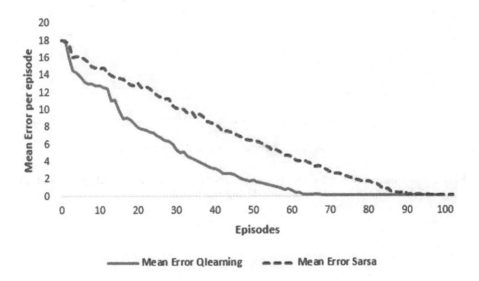

Fig. 1. The mean reward for the 10 gene expression datasets with 20 genes and 20 samples (timepoints) over 100 episodes during the learning process.

4.3   GRN with 50 Genes and 50 Timepoints

In this experiment, we evaluated our proposal in 10 gene expression data matrices, each one containing 50 genes and 50 time samples (timepoints). Differently from the previous experiments, SARSA performed slightly better than QLearning, as shown in Fig. 3. The mean error at the end of the experiments was about 10.0 (considering 10 executions on distinct gene expression data matrices). The minimum error obtained was 8.0 for both Exhaustive Search and RL, but RL could find the minimum error after analyzing 10,421 predictor subsets, while ES had to analyze all triples (combination of 50, 3 to 3 = 19,600 triples).

5   Discussion

First, it is important to notice the challenge of limited number of samples, which is the case for most gene expression data samples available. The lack of samples leads to non-observed or poorly observed predictor instances. This leads the

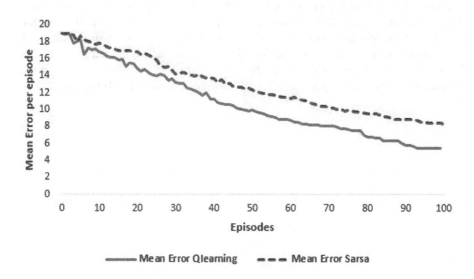

Fig. 2. The mean reward for the 10 gene expression datasets with 20 genes and 50 samples (timepoints) over 100 episodes during the learning process.

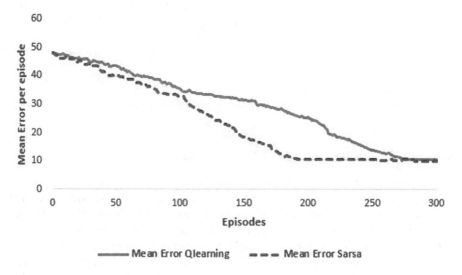

Fig. 3. The mean reward for the 10 gene expression datasets with 50 genes and 50 samples (timepoints) over 300 episodes during the learning process.

error to be almost 0, which could be misleading, since the estimation of the joint probability distributions between the target and the candidate predictor subsets is problematic.

The ES presents the best possible result under the significant search space constrained when considering only subsets with three predictors (triples), while both RL algorithms were able to achieve almost the same results as the ES (the optimal ones for that particular number of predictors), without constraining the search space and taking only a fraction of computing time to find the best solution. As the ES is feasible for pairs or triples of predictors at most, considering real sized networks (with thousands of genes), RL has potential to achieve even better results, since it could retrieve more than 3 predictors per gene. But due to the drawback of excessive memory consuming by our first attempt of implementation of the RLs adapted to GRN inference domain, we still could not consider larger networks, so scalability of RLs remains to be demonstrated in further memory efficient implementations.

Qlearning showed better performance than SARSA for smaller networks (20 genes), but the opposite happened for larger networks (50 genes). Despite scalability challenges, especially regarding excessive memory consuming, we note that both of them are promising and can provide as good results as the ES with few steps and in less time. With some better adaptation of these RL methods to the GRN inference domain, they can be useful for real-sized networks, as they showed their potential to deal with "real world" problems [4, 8, 32].

6   Conclusion

Reinforcement Learning (RL) is a powerful learning paradigm with many potential uses and applications, including GRN inference, a poorly explored domain as far as we know. The first implementation attempt of QLearning and SARSA algorithms for GRN inference provided here shows promigins, but still presents excessive memory consumption which leads to scalability limitations. The previous knowledge the agent has and the way in which it perceives its expected future rewards influences how it learns and the final policy it achieves. Besides, more knowledge about the environment and more samples lead to faster learning process. If an agent acts with a stochastic policy, it will be more uncertain about rewards and choose a safer path. On the other hand, an agent that uses just a greedy approach, expects to always select the best action and will take more direct and riskier actions. Understanding these limitations and sensitivity of RL algorithms for GRN inference is key to comprehend how to properly implement them to deal with real-sized networks.

For further works, other approaches, algorithms and metrics can be explored, including not only information about the topology, but also some information about the dynamics generated by the networks. Other RL apporaches include Options [4], Multiobjective [5] or Multiagent Reinforcement Learning [26], which might accelerate the learning process and also can give weights according to the importance of some genes in the networks. It could be done by selecting the priority genes in a network, and then the RL algorithms would incorporate information about it. Another benefit is that these approaches can split the learning process in smaller parts and learn various parts of the problem simultaneously. This could alleviate the memory consuming, leading to better scalability.

Acknowledgment. We are grateful for the financial support from Coordenação de Aperfeiçoamento de Pessoal de Nível Superior - Brazil (CAPES) - Finance Code 001 and Fundação de Amparo a Pesquisa do Estado de São Paulo (FAPESP), grants 2018/18560-6 and 2018/21934-5. We also thank Felipe Leno da Silva for important technical discussions about RL.

References

1. Akutsu, T., Miyano, S., Kuhara, S., et al.: Identification of genetic networks from a small number of gene expression patterns under the Boolean network model. In: Proceedings of the Pacific Symposium on Biocomputing (PSB), vol. 4, pp. 17–28 (1999)
2. Anastassiou, D.: Computational analysis of the synergy among multiple interacting genes. Mole. Syst. Biol. **3**, 83 (2007)
3. Barrera, J., et al.: Constructing probabilistic genetic networks of Plasmodium falciparum from dynamical expression signals of the intraerythrocytic development cycle. In: McConnell, P., Lin, S.M., Hurban, P. (eds.) Methods of Microarray Data Analysis, pp. 11–26. Springer, Boston, MA (2007). https://doi.org/10.1007/978-0-387-34569-7_2
4. Bonini, R., Da Silva, F.L., Glatt, R., Spina, E., Costa, A.H.R.: A framework to discover and reuse object-oriented options in reinforcement learning. In: 2018 7th Brazilian Conference on Intelligent Systems (BRACIS), pp. 109–114. IEEE (2018)
5. Bonini, R.C., Silva, F.L., Spina, E., Costa, A.H.R.: Using options to accelerate learning of new tasks according to human preferences. In: AAAI Workshop Human-Machine Collaborative Learning, pp. 1–8 (2017)
6. Brazhnik, P., Fuente, A., Mendes, P.: Gene networks: how to put the function in genomics. Trends Biotechnol. **20**(11), 467–472 (2002)
7. Cover, T.M., Van-Campenhout, J.M.: On the possible orderings in the measurement selection problem. IEEE Trans. Syst. Man Cybern. **7**(9), 657–661 (1977)
8. Da Silva, F.L., Nishida, C.E., Roijers, D.M., Costa, A.H.R.: Coordination of electric vehicle charging through multiagent reinforcement learning. IEEE Trans. Smart Grid **11**(3), 2347–2356 (2019)
9. De Jong, H.: Modeling and simulation of genetic regulatory systems: a literature review. J. Comput. Biol. **9**(1), 67–103 (2002)
10. D'haeseleer, P., Liang, S., Somgyi, R.: Tutorial: gene expression data analysis and modeling. In: Pacific Symposium on Biocomputing. Hawaii, January 1999
11. Dougherty, E.R., Xiao, Y.: Design of probabilistic Boolean networks under the requirement of contextual data consistency. IEEE Trans. Signal Process. **54**(9), 3603–3613 (2006)
12. Eberwine, J., Sul, J., Bartfai, T., Kim, J.: The promise of single-cell sequencing. Nat. Methods **11**, 25–27 (2014)
13. Erdös, P., Rényi, A.: On random graphs. Publ. Math. Debrecen **6**, 290–297 (1959)
14. Hecker, M., Lambeck, S., Toepfer, S., van Someren, E., Guthke, R.: Gene regulatory network inference: data integration in dynamic models-a review. Biosystems **96**, 86–103 (2009)

15. Jacomini, R.S., Martins-Jr, D.C., Silva, F.L., Costa, A.H.R.: GeNICE: a novel framework for gene network inference by clustering, exhaustive search, and multi-variate analysis. J. Comput. Biol. **24**(8), 809–830 (2017)

16. Jimenez, R.D., Martins-Jr, D.C., Santos, C.S.: One genetic algorithm per gene to infer gene networks from expression data. Netw. Modeling Anal. Health Inform. Bioinform. **4**, 1–22 (2015)

17. Kauffman, S.A.: Homeostasis and differentiation in random genetic control networks. Nature **224**(215), 177–178 (1969)

18. Liang, S., Fuhrman, S., Somogyi, R.: Reveal, a general reverse engineering algorithm for inference of genetic network architectures. In: Pacific Symposium on Biocomputing, vol. 3, pp. 18–29 (1998)

19. Lopes, F.M., Martins-Jr, D.C., Barrera, J., Cesar-Jr, R.M.: A feature selection technique for inference of graphs from their known topological properties: revealing scale-free gene regulatory networks. Inf. Sci. **272**, 1–15 (2014)

20. Marbach, D., Prill, R.J., Schaffter, T., Mattiussi, C., Floreano, D., Stolovitzky, G.: Revealing strengths and weaknesses of methods for gene network inference. Proc. Natl. Acad. Sci. **107**(14), 6286–6291 (2010)

21. Martins-Jr, D.C., Braga-Neto, U., Hashimoto, R.F., Dougherty, E.R., Bittner, M.L.: Intrinsically multivariate predictive genes. IEEE J. Sel. Top. Signal Process. **2**(3), 424–439 (2008)

22. Nam, D., Seo, S., Kim, S.: An efficient top-down search algorithm for learning Boolean networks of gene expression. Mach. Learn. **65**, 229–245 (2006)

23. Pratapa, A., Jalihal, A.P., Law, J.N., Bharadwaj, A., Murali, T.: Benchmarking algorithms for gene regulatory network inference from single-cell transcriptomic data. Nat. Methods **17**(2), 147–154 (2020)

24. Shalon, D., Smith, S.J., Brown, P.O.: A DNA microarray system for analyzing complex DNA samples using two-color fluorescent probe hybridization. Genome Res. **6**, 639–45 (1996)

25. Shmulevich, I., Dougherty, E.R., Kim, S., Zhang, W.: Probabilistic Boolean networks: a rule-based uncertainty model for gene regulatory networks. Bioinformatics **18**(2), 261–274 (2002)

26. Silva, F.L., Taylor, M.E., Costa, A.H.R.: Autonomously reusing knowledge in multiagent reinforcement learning. In: IJCAI (2018)

27. Snoep, J.L., Westerhoff, H.V.: From isolation to integration, a systems biology approach for building the silicon cell. Top. Curr. Genet. **13**, 13–30 (2005)

28. Sutton, R.S., Barto, A.G.: Reinforcement Learning: An Introduction, 1st edn. MIT Press, Cambridge (1998)

29. Velculescu, V.E., Zhang, L., Vogelstein, B., Kinzler, K.W.: Serial analysis of gene expression. Science **270**, 484–487 (1995)

30. Wang, Z., Gerstein, M., Snyder, M.: RNA-SEQ: a revolutionary tool for transcriptomics. Nat. Rev. Genet. **10**(1), 57–63 (2009)

31. Watkins, C.J., Dayan, P.: Q-learning. Mach. Learn. **8**(3), 279–292 (1992)

32. Yerudkar, A., Chatzaroulas, E., Del Vecchio, C., Moschoyiannis, S.: Sampled-data control of probabilistic Boolean control networks: a deep reinforcement learning approach. Inf. Sci. **619**, 374–389 (2023)

33. Zhang, Y., Chang, X., Liu, X.: Inference of gene regulatory networks using pseudo-time series data. Bioinformatics **37**(16), 2423–2431 (2021)

Exploring Identifiability in Hybrid Models of Cell Signaling Pathways

Ronaldo N. Sousa[1,2]([envelope]) [iD], Cristiano G. S. Campos[4] [iD], Willian Wang[1,2] [iD], Ronaldo F. Hashimoto[1] [iD], Hugo A. Armelin[2,3] [iD], and Marcelo S. Reis[2,4] [iD]

[1] Institute of Mathematics and Statistics, University of São Paulo, São Paulo, Brazil
`ronogue.rn@ime.usp.br`
[2] CeTICS, Cell Cycle lab, Butantan Institute, São Paulo, Brazil
[3] Institute of Chemistry, University of São Paulo, São Paulo, Brazil
[4] Recod lab, Institute of Computing, University of Campinas, Campinas, Brazil
`cristiano.campos@students.ic.unicamp.br`, `msreis@ic.unicamp.br`

Abstract. Various processes, including growth, proliferation, migration, and death, mediate the activity of a cell. To better understand these processes, dynamic modeling can be a helpful tool. First-principle modeling provides interpretability, while data-driven modeling can offer predictive performance using models such as neural network, however at the expense of the understanding of the underlying biological processes. A hybrid model that combines both approaches might mitigate the limitations of each of them alone; nevertheless, to this end one needs to tackle issues such as model calibration and identifiability. In this paper, we report a methodology to address these challenges that makes use of a universal differential equation (UDE)-based hybrid modeling, were a partially known, ODE-based, first-principle model is combined with a feedforward neural network-based, data-driven model. We used a synthetic signaling network composed of 38 chemical species and 51 reactions to generate simulated time series for those species, and then defined twelve of those reactions as a partially known first-principle model. A UDE system was defined with this latter and it was calibrated with the data simulated with the whole network. Initial results showed that this approach could identify the missing communication of the partially-known first-principle model with the remainder of the network. Therefore, we expect that this type of hybrid modeling might become a powerful tool to assist in the investigation of underlying mechanisms in cellular systems.

Keywords: Scientific Machine Learning · First-principle Modeling · Universal Differential Equation · Inverse Problem · Cell Signaling Pathway

1 Introduction

The activity of a cell is mediated by various processes such as growth, proliferation, migration, death, and others. These processes are orchestrated by messages transmitted through the so-called cell signaling pathways. Such pathways

© The Author(s), under exclusive license to Springer Nature Switzerland AG 2023
M. S. Reis and R. C. de Melo-Minardi (Eds.): BSB 2023, LNBI 13954, pp. 148–159, 2023.
https://doi.org/10.1007/978-3-031-42715-2_14

are composed of a cascade of chemical reactions (e.g., enzymatic reactions) in which products of one reaction are used as substrates for another. A healthy cell requires its cellular processes to be adequately orchestrated. Dysregulations in cell signaling pathways are involved in the pathophysiology of many diseases, including cancer [8,9].

Cell signaling pathways are non-linear systems whose dynamics depends on concentration changes of the involved chemical species (e.g., proteins) over a given period. Due to the intrinsic non-linearity of those systems, human intuition is not enough to infer the behavior of a particular signaling pathway when its initial conditions are modified (i.e., how a specific signaling pathway behaves from different cellular stimuli, such as the addition of different compounds to the cell culture). Therefore, mathematical modeling of cell signaling pathways is an important tool in studying the mechanisms behind cellular processes. Two widely used approaches for modeling of cell signaling pathways are first-principle and data-driven.

First-principle modeling involves describing signaling pathways based on the underlying physical-chemical principles of their activity. In this approach, ordinary differential equations (ODEs) describe the signaling pathways where each ODE translates into how the involved chemical species are consumed or produced in terms of kinetic laws [5,18]. First-principle modeling offers the benefit of interpretability since it explicitly correlates the input and outputs by employing the laws of chemical kinetics. On the other hand, just a small set of reactions and chemical species can be modeled since the degree of freedom of the model increases as the number of components in the model increases (e.g., the number of reactions and chemical species). Furthermore, sometimes one only partially knows the underlying cellular process mechanism, making it challenging to develop a precise first-principle model [1,5,13].

Data-driven modeling employs machine learning techniques to predict the behavior of a signaling pathway from experimental measurements of chemical species present in the cell. Opposite to the first-principle modeling, data-driven modeling relies solely on produced data. Among the techniques commonly used for this purpose are linear models, tree-based models, neural networks, and ensemble methods, with varied results [6]. Data-driven models are often less interpretable than first-principle ones, especially when one uses more powerful and complex models such as neural networks.

Therefore, it would be interesting to combine the interpretability of first-principle modeling with the power of data-driven modeling. To this end, one approach is to construct a hybrid model where the first-principle and data-driven models are integrated for modeling of cell signaling pathways. More precisely, the first-principle model is used to express what is known about the underlying process, whereas the data-driven model is employed to learn what is missing from the first-principle model [12,13]. However, even when the first-principle model has its parameters completely identified, the associated data-driven model still has to be learned (e.g., weights values in a neural network). Moreover, the identifiability of such hybrid model is still underexplored in the context of cell signaling pathways.

This work aims to explore the identifiability of the data-driven part of hybrid models of cell signaling pathways. This paper consists of five sections, including this introduction. Section 2 is dedicated to discuss the related works within the context of our own work. In Sect. 3, we describe the proposed methodology, including the presentation of the used toy model, and also the experimental setup. In Sect. 4, we present the initial results obtained with the proposed methodology. Finally, In Sect. 5, we summarize the findings of this paper and outline the next steps of this research.

2   Related Works

One of the first approaches to deal with the intrinsic incompleteness of signaling pathways modeled by ODEs was proposed in 2017 by Engelhardt and colleagues [4]. In that work, the authors proposed a Bayesian approach to estimate latent variables that would be missing from the system. However, the proposed method, called BDEN, only uses information from proteins present in the first-principle part of the model and does not allow the incorporation of existing prior information for the remainder of the cell [4]. To address the limitations of signaling pathways modeled only with ODEs, in 2021 Glass and colleagues proposed the usage of delay differential equations, with promising results [7]. Another pursued path is data-driven modeling, using machine learning; for instance, in 2021, Gabor and colleagues described the outcome of a competition in which different machine learning techniques (linear models, decision trees, neural networks, ensemble methods, among others) were applied to large-scale experimental data produced for different human breast cancer cell lines subjected to different treatments, achieving different levels of performance [6].

There has been growing interest in a hybrid modeling approach combining first-principle models with data-driven models in recent years. One example is the development of a hybrid model in 2020, which aimed to model a signaling pathway that was only partially observed in the network [12]; however, the design of such a model is overly complicated and with some ad hoc procedures [12]. In 2021, a mathematical object called Universal Differential Equations (UDEs) was introduced, which is a tool for coupling mathematical descriptions of natural laws with data-driven machine learning approaches [17]. UDEs were used by Bangi and colleagues in 2022 for the modeling of beta-carotene production using yeast, with superior performance in comparison to the classical approach [3]. More recently, in 2023, Santana and colleagues discussed using a hybrid modeling approach to model the sorption uptake kinetics systematically and efficiently, which resulted in a well-fitted hybrid model. In the end, sparsed and symbolic regression was used to reconstruct the sorption uptake kinetics showing that the proposed approach holds promising potential to discover sorption kinetic law structures [19].

3   Methodology

The methodology of this work consists in the usage of a synthetic model that represents a complete first-principle model of a cell signaling network, composed of 38 chemical species and 51 chemical reactions (Fig. 1). That model is complete in the sense that the model is an isolated system. We use that model to generate a large amount of time series data of chemical species that belongs to that system. In the sequence, we choose a subset composed of twelve chemical reactions of that system to represent a cell signaling pathway (Fig. 1, in blue), whose communication with the remainder of the network will be inferred using the produced data and a hybrid model (i.e., a model that combines a first-principle model with a data-driven model).

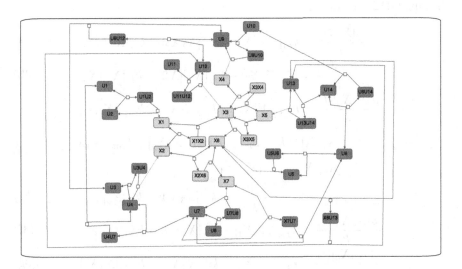

Fig. 1. Diagram in Systems Biology Graphical Notation (SBGN) [2,11] presenting the synthetic cell signaling network used in this work. Nodes and edges represent, respectively, chemical species (e.g., proteins) and reactions. The sets of blue and purple nodes symbolize, respectively, a cell signaling pathway and species outside of that pathway.

The first-principle model of the cell signaling pathway (subset of the complete signaling network, nodes in blue in Fig. 1) has the following set of reactions:

$$X_1 + X_2 \underset{k_{r_1}}{\overset{k_{f_1}}{\rightleftharpoons}} X_1X_2 \xrightarrow{k_{cat_1}} X_1 + X_3 \tag{1a}$$

$$X_3 + X_4 \underset{k_{r_2}}{\overset{k_{f_2}}{\rightleftharpoons}} X_3X_4 \xrightarrow{k_{cat_2}} X_3 + X_5 \tag{1b}$$

$$X_3 + X_5 \underset{k_{r_3}}{\overset{k_{f_3}}{\rightleftharpoons}} X_3X_5 \xrightarrow{k_{cat_3}} X_3 + X_6 \tag{1c}$$

$$X_2 + X_6 \underset{k_{r_4}}{\overset{k_{f_4}}{\rightleftharpoons}} X_2X_6 \xrightarrow{k_{cat_4}} X_6 + X_7. \tag{1d}$$

Those reactions were then transcribed to a set of ordinary differential equations. For each chemical species, an ODE was defined by the inclusion of terms that describes its production or consumption in a given first or second-order reaction. For instance, X_1 is consumed in a second-order reaction and is produced in two first-order reaction (Eq. 1a), thus resulting the following ODE:

$$\frac{d[X_1](t)}{dt} = k_{cat_1}[X_1X_2] + k_{r_1}[X_1X_2] - k_{f_1}[X1][X2]. \tag{2}$$

ODEs for the other species were generated in the same way, thus yielding the following ODE system:

$$\frac{d[X_1](t)}{dt} = k_{cat_1}[X_1X_2] + k_{r_1}[X_1X_2] - k_{f_1}[X1][X2] \tag{3a}$$

$$\frac{d[X_2](t)}{dt} = k_{r_1}[X_1X_2] - k_{f_1}[X1][X2] - k_{f_4}[X_2][X_6] \tag{3b}$$

$$\frac{d[X_1X_2](t)}{dt} = k_{f_1}[X1][X2] - k_{r_1}[X_1X_2] - k_{cat_1}[X_1X_2] \tag{3c}$$

$$\frac{d[X_3](t)}{dt} = k_{r_2}[X_3X_4] + k_{r_3}[X_3X_5] + k_{cat_2}[X_3X_4] + k_{cat_3}[X_3X_5]$$
$$- k_{f_3}[X_3][X_5] - k_{f_2}[X_3][X_4] \tag{3d}$$

$$\frac{d[X_4](t)}{dt} = k_{r_2}[X_3X_4] - k_{f_2}[X_3][X_4] \tag{3e}$$

$$\frac{d[X_3X_4](t)}{dt} = k_{f_2}[X_3][X_4] - k_{r_2}[X_3X_4] - k_{cat_2}[X_3X_4] \tag{3f}$$

$$\frac{d[X_3X_5](t)}{dt} = k_{f_3}[X_3][X_5] - k_{r_3}[X_3X_5] - k_{cat_3}[X_3X_5] \tag{3g}$$

$$\frac{d[X_5](t)}{dt} = k_{r_3}[X_3X_5] - k_{f_3}[X_3][X_5] \tag{3h}$$

$$\frac{d[X_6](t)}{dt} = k_{r_4}[X_2X6] - k_{f_4}[X_2][X_6] + k_{cat_4}[X_2X_6] \tag{3i}$$

$$\frac{d[X_5X_6](t)}{dt} = k_{f_4}[X_2][X_6] - k_{r_4}[X_2X6] - k_{cat_4}[X_2X_6] \tag{3j}$$

$$\frac{d[X_7](t)}{dt} = k_{cat_4}[X_2X_6]. \tag{3k}$$

To conduct the experiments, we assumed that all rate constants of the first-principle model are known (Table 1). These constants were carefully selected to ensure the generation of diverse and comprehensive data. To achieve this, we systematically simulated the model with various rate constants, evaluating and selecting those that best met the criteria.

Using synthetic data instead of real measured data is justified due to its ability to generate diverse data with different conditions and its lack of limitations and constraints associated with real data collection, such as experimental uncertainties, limited number of samples, and noise. Synthetic data provides flexibility in assessing the identifiability of hybrid modeling in inferring missing signals in cell signaling pathway models. However, future work should include employing the methodology with real measured data, thereby further enhancing the validity and applicability of the findings.

Table 1. Rate constants values used to parameterize the first-principle model of the cell signaling pathway. This set of parameters corresponds to θ in Eq. 4.

k_{f_1}	k_{r_1}	k_{cat_1}
0.0150	0.1000	0.0030

k_{f_2}	k_{r_2}	k_{cat_2}
0.0990	0.1150	0.0850

k_{f_3}	k_{r_3}	k_{cat_3}
0.0890	0.0500	0.1500

k_{f_4}	k_{r_4}	k_{cat_4}
0.2500	0.4325	0.0150

3.1   Hybrid Model

Modeling a cell signaling pathway using an ODE-based, first-principle model does not take into account the communication with the remainder of the complete network. This can lead to a model whose dynamics diverges from the actual cell signaling pathway, since it would be modeled as it was an isolated system, which is not the case. Therefore we tackle this problem using a hybrid model, namely with a set of universal differential equations (UDEs) [17]. UDEs combine an ODE-based, first-principle model with a data-driven model (a feedforward neural network):

$$\dot{\mathbf{x}} = f_\theta(\mathbf{x}) + U_\mathbf{w}(\mathbf{x}), \tag{4}$$

where the $f : \mathbb{R}^n_{\geq 0} \to \mathbb{R}^n_{\geq 0}$ is a function parametrized by θ that maps a vector of species concentration $\mathbf{x}$ to another vector of same type, and U is a neural network with weight vector $\mathbf{w}$. In the case of our hybrid model, f is defined using the ODE system of Eq. 3a–3k.

For our experiments, the neural network U in Eq. 4 was set with an input layer with eleven nodes, one hidden layer with seven nodes, an output layer with seven nodes (we assume that four species are not produced throughout the experiment). The sigmoid was used as the activation function in the hidden layer. The hybrid model was implemented using the Scientific Machine Learning Ecosystem [14–16], which is coded in Julia. The source code of our experiments is open and free, under the MIT license, and can be accessed at github.com/Dynamic-Systems-Biology/BSB-2023-Hybrid-modeling.

3.2   Model Training and Assessment

The model described in Fig. 1 was simulated 30 times; for each simulation, the initial condition vector was sampled from a uniform distribution on the interval $[0, 13)$; the considered sampled initial conditions are showed in Tables 2, 3 and 4. Finally, those produced time series were split into training, validation and test sets of equal size. The parameter vector $\mathbf{w}$ was optimized with three runs using the ADAM optimizer [10] with the sequence of learning rates $(0.1, 0.05, 0.001)$. After that, the BFGS optimizer was used [20]. The validation set was used to make the early stopping decision to avoid overfitting. The criterion to employ the early stopping was to verify whether the loss value calculated on the validation set increased after 100 iterations. The loss function was defined as:

$$\ell(\hat{\boldsymbol{X}}, \boldsymbol{X}, \mathbf{w}) = \frac{1}{N} \sum_{i=1}^{I} \sum_{j=1}^{J} |\hat{\boldsymbol{X}}_{ij} - \boldsymbol{X}_{ij}| + \lambda \sum_{k=0}^{W} \mathbf{w}_k^2, \qquad (5)$$

where N is the number of data points, $\boldsymbol{X}$ $(\hat{\boldsymbol{X}})$ is a $I \times J$ (11×101) matrix containing the simulated (predicted) time series for each chemical species, and $\mathbf{w}$ is the U parameter vector. The regularization constant λ was chosen to be 10^{-3}. To evaluate the robustness of the hybrid model, we used the mean absolute error (MAE):

$$MAE(\hat{\boldsymbol{X}}, \boldsymbol{X}) = \frac{1}{N} \sum_{i=1}^{I} \sum_{j=1}^{J} |\hat{\boldsymbol{X}}_{ij} - \boldsymbol{X}_{ij}|, \qquad (6)$$

and also the symmetric mean absolute percentage error (SMAPE):

$$SMAPE(\hat{\boldsymbol{X}}, \boldsymbol{X}) = \frac{100}{N} \sum_{i=1}^{I} \sum_{j=1}^{J} \frac{|\hat{\boldsymbol{X}}_{ij} - \boldsymbol{X}_{ij}|}{\hat{\boldsymbol{X}}_{ij} + \boldsymbol{X}_{ij}}. \qquad (7)$$

For both metrics above, a lower value indicates better model performance, with zero being an optimal performance.

Table 2. Initial conditions vectors used to generate the training dataset. The values are truncated to three decimal digits.

u_0^1	2.923	3.742	1.352	6.184	5.416	6.778	11.810	1.336	8.715	9.820	8.437
u_0^2	3.160	4.733	0.855	5.329	7.089	4.357	7.383	10.642	2.872	2.016	5.657
u_0^3	10.552	12.849	10.499	12.611	1.820	6.622	0.763	0.055	12.670	7.243	6.037
u_0^4	8.841	11.367	12.012	12.081	8.128	9.584	3.475	10.458	5.764	9.520	5.055
u_0^5	9.580	2.852	4.642	9.367	7.467	5.073	4.058	9.638	10.744	12.921	6.366
u_0^6	9.511	9.252	0.684	9.550	8.362	12.869	10.673	6.421	8.312	4.092	6.363
u_0^7	9.049	1.887	9.212	0.786	11.772	11.929	1.471	4.334	7.313	0.851	5.030
u_0^8	11.886	7.346	10.572	2.916	3.579	4.668	9.483	4.105	10.673	1.891	12.930
u_0^9	0.749	7.558	9.563	3.703	8.415	6.052	3.767	6.930	6.409	2.828	8.028
u_0^{10}	1.463	4.788	4.477	0.736	1.570	2.334	4.963	10.596	3.148	10.657	8.709

4 Results

During the training of the hybrid model, the loss function was plotted to compare how it behaves in the training and in the validation datasets. As shown in Fig. 2, the chosen parameters resulted in the loss function value of 54.91 on the validation dataset.

To assess how the predictions from the model varied for each initial condition across the datasets, we computed the SMAPE and MAE values for each initial condition in the training, validation and test sets (Tables 5 and 6). The minimum SMAPE and MAE values on the test dataset were 12.74 and 2.10, respectively, while the maximum values were 33.24 and 6.64.

We also evaluated the performance of a partially known first-principle model without a data-driven component, that is, an implementation of the ODE system of Eq. 3a–3k. We used that model to generate simulations with the initial conditions of the 10 test time series and then calculated SMAPE and MAE values (Tables 5 and 6).

Table 3. Initial conditions vectors used to generate the validation dataset. The values are truncated to three decimal digits.

u_0^1	6.850	12.970	7.162	10.430	9.818	0.939	11.753	2.702	5.651	6.048	11.895
u_0^2	7.568	12.695	5.476	6.402	10.944	8.996	5.366	8.677	10.458	0.291	8.746
u_0^3	11.454	4.197	3.840	8.014	5.317	12.370	7.010	11.929	12.444	11.511	7.423
u_0^4	2.331	11.425	4.955	6.098	4.924	6.174	11.739	11.432	3.773	11.092	2.302
u_0^5	0.294	12.573	11.047	1.981	1.970	3.625	2.695	5.588	3.888	11.917	10.714
u_0^6	10.789	10.179	6.083	1.428	12.377	4.338	1.325	1.497	1.282	11.621	11.498
u_0^7	8.446	1.565	7.784	0.273	0.738	5.448	8.503	6.590	9.284	8.628	4.626
u_0^8	6.422	9.774	12.555	12.320	11.447	4.377	11.386	7.304	9.919	4.855	5.524
u_0^9	0.145	7.104	1.283	2.676	11.403	10.758	8.077	1.004	6.945	10.115	12.979
u_0^{10}	8.864	10.12	4.193	0.476	1.983	12.988	5.162	5.033	7.373	8.848	2.452

Table 4. Initial conditions vectors used to generate the test dataset. The values are truncated to three decimal digits.

u_0^1	11.879	4.637	2.379	2.687	6.409	5.214	5.540	10.286	12.693	1.490	5.791
u_0^2	8.415	4.919	0.566	0.987	6.568	10.816	0.725	6.562	5.852	8.111	7.152
u_0^3	1.347	5.528	1.277	1.753	12.682	4.502	9.385	0.744	0.958	7.303	4.366
u_0^4	5.695	5.321	2.487	10.535	9.703	12.073	9.179	0.125	0.246	8.342	7.701
u_0^5	2.501	9.231	0.260	8.329	7.844	5.686	0.136	3.298	10.898	7.923	9.751
u_0^6	0.295	5.092	7.221	2.668	3.000	3.074	3.765	6.705	10.842	10.876	0.130
u_0^7	11.259	6.948	10.358	11.984	10.365	6.930	11.746	0.151	9.053	6.123	1.373
u_0^8	1.799	2.410	2.207	11.627	1.592	8.172	3.196	3.647	10.349	6.833	0.599
u_0^9	4.892	2.003	0.903	7.908	7.938	9.758	2.323	11.508	1.042	3.829	6.756
u_0^{10}	3.491	4.812	4.737	8.514	8.833	4.911	12.051	5.260	6.625	9.282	3.063

Finally, we also made a qualitative verification on the best and worst fittings, that is, we analyzed how far lied the predicted curves from the measured ones. In Figs. 3 and 4, we show that analysis for the hybrid model best and worst predictions, respectively.

Table 5. SMAPE values for each of the 10 initial conditions. The last row is the results when only the partially known first-principle model is used for prediction.

	u_0^1	u_0^2	u_0^3	u_0^4	u_0^5	u_0^6	u_0^7	u_0^8	u_0^9	u_0^{10}	Mean
Training	**18.6**	42.1	23.4	**12.2**	23.7	21.5	27.7	**16.1**	**16.3**	**17.8**	**21.9**
Validation	30.6	**22.3**	**19.7**	21.8	36.3	**14.1**	**20.0**	43.9	23.6	18.4	25.1
Test	30.0	23.9	29.2	12.7	**25.9**	19.0	27.8	33.2	17.2	27.2	24.6
First-principle only	68.0	62.0	55.7	55.5	49.6	61.7	63.3	54.6	55.6	59.2	58.5

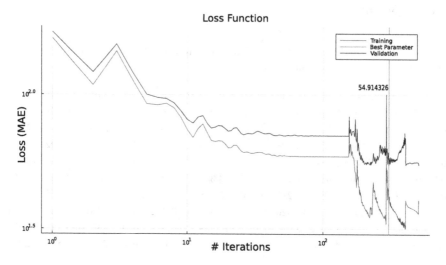

Fig. 2. The loss function value per iteration. The black vertical line marks the point at which the parameter values result in the minimum loss on the validation set.

Table 6. MAE values for each of the 10 initial conditions. The last row is the results when only the partially known first-principle model is used for prediction.

	u_0^1	u_0^2	u_0^3	u_0^4	u_0^5	u_0^6	u_0^7	u_0^8	u_0^9	u_0^{10}	Mean
Training	**3.7**	11.5	**3.4**	2.3	**3.3**	4.3	4.3	**2.6**	**2.4**	**2.2**	**4.0**
Validation	6.0	5.0	4.1	5.2	8.4	**2.8**	**2.6**	12.0	4.3	4.0	5.4
Test	6.0	**4.6**	5.6	**2.1**	3.9	3.1	5.5	6.6	3.9	5.5	4.7
First-principle only	15.2	14.2	9.4	10.4	11.2	9.0	12.6	11.1	12.2	11.0	11.6

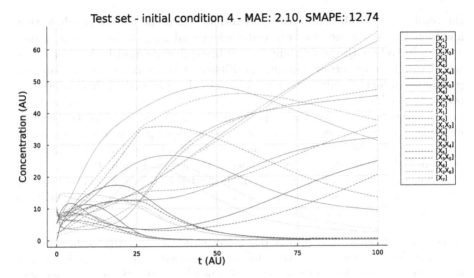

Fig. 3. The best prediction of the hybrid model with SMAPE and MAE metrics of 12.74 and 2.10, respectively.

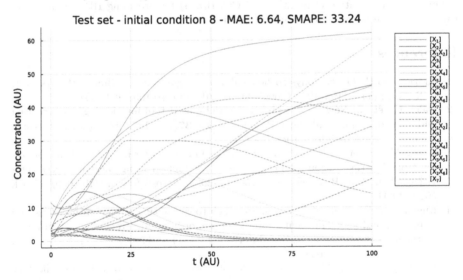

Fig. 4. The worst prediction of the hybrid model with SMAPE and MAE metrics of 33.24 and 6.64, respectively.

5   Final Remarks

Our results presented in this report suggest that the hybrid modeling approach for cell signaling pathways is more effective than the partially known first-principle model alone (Tables 5 and 6). These findings provide promising insights into the potential effectiveness of hybrid modeling in this context. By comparing the results of the validation, test, and training datasets, we can conclude that identifiability was achievable, especially when analyzing the predictions generated exclusively with the partially known first-principle model. However, these initial results have certain limitations. Firstly, it is important to validate the possibility of identifiability other models than the one used in this paper. Secondly, we need to experiment with varying sizes of training, validation, and test datasets to determine whether the performance metrics improve as the data size increases. Thirdly, it is essential to optimize the hyper-parameters of the neural network and the training process, such as the number of layers and learning rates, to further improve the performance of the model. Finally, there is a problem that arises when the parameters of the first-principle model is partially or totally unknown; for that case, we need to find a way to infer them at same time the neural network is trained.

Acknowledgements. The authors thank Juliane Liepe (Max Planck Institute for Multidisciplinary Sciences, Germany) for her intellectual input in an earlier project in the same research line. This work was supported by CAPES, BECAS Santander and by grants 13/07467-1, 15/22308-2, 19/21619-5, 19/24580-2, and 21/04355-4, São Paulo Research Foundation (FAPESP). The authors also acknowledge the National Laboratory for Scientific Computing (LNCC, Brazil) for providing HPC resources of the SDumont supercomputer (sdumont.lncc.br).

References

1. Aldridge, B.B.: Others: physicochemical modelling of cell signalling pathways. Nat. Cell Biol. **8**(11), 1195–1203 (2006). https://doi.org/10.1038/ncb1497
2. Balci, H., et al.: Newt: a comprehensive web-based tool for viewing, constructing and analyzing biological maps. Bioinformatics. **37**(10), 1475–1477 (2020). https://doi.org/10.1093/bioinformatics/btaa850
3. Bangi, M.S.F., et al.: Physics-informed neural networks for hybrid modeling of lab-scale batch fermentation for beta-carotene production using saccharomyces cerevisiae. Chem. Eng. Res. Des. **179**, 415–423 (2022). https://doi.org/10.1016/j.cherd.2022.01.041
4. Engelhardt, B., et al.: A Bayesian approach to estimating hidden variables as well as missing and wrong molecular interactions in ordinary differential equation-based mathematical models. J. R. Soc. Interface **14**(131), 20170332 (2017). https://doi.org/10.1098/rsif.2017.0332
5. Fröhlich, F., et al.: Mechanistic model of MAPK signaling reveals how allostery and rewiring contribute to drug resistance. Mol. Syst. Biol. **19**(2), e10988 (2023). https://doi.org/10.15252/msb.202210988
6. Gabor, A., et al.: Cell-to-cell and type-to-type heterogeneity of signaling networks: insights from the crowd. Mol. Syst. Biol. **17**(10), e10402 (2021). https://doi.org/10.15252/msb.202110402
7. Glass, D.S., et al.: Nonlinear delay differential equations and their application to modeling biological network motifs. Nat. Commun. **12**(1), 1788 (2021). https://doi.org/10.1038/s41467-021-21700-8
8. Hidalgo, M.R., et al.: Models of cell signaling uncover molecular mechanisms of high-risk neuroblastoma and predict disease outcome. Biol. Direct **13**(1), 16 (2018). https://doi.org/10.1186/s13062-018-0219-4

9. Joo, J.D.: The use of intra-cellular signaling pathways in anesthesiology and pain medicine field. Korean J. Anesthesiol. **57**(3), 277–283 (2009). https://doi.org/10.4097/kjae.2009.57.3.277

10. Kingma, D.P., Ba, J.: Adam: A method for stochastic optimization. ArXiv Preprint ArXiv:1412.6980 (2014). https://doi.org/10.48550/arXiv.1412.6980

11. Le Novère, N., et al.: The systems biology graphical notation. Nat. Biotechnol. **27**(8), 735–741 (2009). https://doi.org/10.1038/nbt.1558

12. Lee, D., et al.: Development of a hybrid model for a partially known intracellular signaling pathway through correction term estimation and neural network modeling. PLoS Comput. Biol. **16**(12), 1–31 (2020). https://doi.org/10.1371/journal.pcbi.1008472

13. Lee, D., et al.: A hybrid mechanistic data-driven approach for modeling uncertain intracellular signaling pathways. In: 2021 American Control Conference (ACC), pp. 1903–1908 (2021). https://doi.org/10.23919/ACC50511.2021.9483352

14. Ma, Y., et al.: A comparison of automatic differentiation and continuous sensitivity analysis for derivatives of differential equation solutions. In: 2021 IEEE High Performance Extreme Computing Conference (HPEC), pp. 1–9 (2021). https://doi.org/10.1109/HPEC49654.2021.9622796

15. Pal, A.: Lux: Explicit parameterization of deep neural networks in Julia (2022). https://github.com/avik-pal/Lux.jl/

16. Rackauckas, C., Nie, Q.: Differentialequations.jl-a performant and feature-rich ecosystem for solving differential equations in Julia. J. Open Res. Softw. **5**(1), 15 (2017). https://doi.org/10.5334/jors.151

17. Rackauckas, C., et al.: Universal differential equations for scientific machine learning (2021). https://doi.org/10.48550/arXiv.2001.04385

18. Reis, M.S., et al.: An interdisciplinary approach for designing kinetic models of the RAS/MAPK signaling pathway. In: Methods in Molecular Biology Special Edition on Kinase Signaling Networks, pp. 455–474. Humana Press, New York (2017). https://doi.org/10.1007/978-1-4939-7154-1_28

19. Santana, V.V., et al.: Efficient hybrid modeling and sorption model discovery for non-linear advection-diffusion-sorption systems: a systematic scientific machine learning approach. ArXiv Preprint ArXiv:2303.13555 (2023). https://doi.org/10.48550/arXiv.2303.13555

20. Scheithauer, G., Nocedal, J., Wright, S.J.: Numerical Optimization. Springer Series In Operations Research and Financial Engineering, 2nd edn. Springer, New York (1999)

Author Index

M. S. Reis and R. C. de Melo-Minardi (Eds.): BSB 2023, LNBI 13954, pp. 161–162, 2023.
https://doi.org/10.1007/978-3-031-42715-2

Printed in the United States
by Baker & Taylor Publisher Services

Printed in the United States
by Baker & Taylor Publisher Services